To Cara, who rekindled my interest in law and politics

Constitutional Calculus

Constitutional Calculus

The Math of Justice and
the Myth of Common Sense

Jeff Suzuki

Johns Hopkins University Press
Baltimore

Johns Hopkins University Press
2715 North Charles Street
Baltimore, Maryland 21218-4363
www.press.jhu.edu

Library of Congress Cataloging-in-Publication Data

Suzuki, Jeff.
    Constitutional calculus : the math of justice and the myth of common
sense / by Jeff Suzuki.
        pages cm
    Includes bibliographical references and index.
    ISBN 978-1-4214-1595-6 (hardcover : alk. paper) — ISBN
978-1-4214-1596-3 (electronic) — ISBN 1-4214-1595-X (hardcover : alk.
paper) — ISBN 1-4214-1596-8 (electronic)   1. Representative government and representation—United States.   2. Probabilities—United
States.   3. Mathematical statistics—United States.   4. Voting—United
States.   5. Social justice—United States.   I. Title.
    JK1726.S89 2015
    320.97301'1—dc23   2014014600

A catalog record for this book is available from the British Library.

*Special discounts are available for bulk purchases of this book. For more
information, please contact Special Sales at 410-516-6936 or specialsales@
press.jhu.edu.*

Contents

Acknowledgments

I'd like to thank Stephen Brams for reviewing an early version of the manuscript and pointing out some things I'd missed the first time around. I'd also like to thank my wife, Jacqui, whose constant requests for clarification helped make this a much better work.

Constitutional Calculus

Condorcet's Dream

The application of the arithmetic of combinations and probabilities to [moral and political] sciences, promises an improvement by so much the more considerable, as it is the only means of giving to their results an almost mathematical precision, and of appreciating their degree of certainty or probability.

—MARQUIS DE CONDORCET (1743–1794),
*SKETCH FOR A HISTORICAL PICTURE
OF THE PROGRESS OF THE HUMAN MIND*

A mathematical theory begins with explicit assumptions, known as axioms, and follows their logical consequences. Thus plane geometry begins with axioms like "Two points define a unique straight line" and arrives at statements like "In a right triangle, the squares on the two sides are together equal to the square of the hypotenuse." What we or our forebears believed to be true is irrelevant: mathematics requires us to abandon tradition and common sense, and use only logic.

Human societies, on the other hand, are usually based on tradition and common sense. For millennia, it was common sense that the vast majority of the population was incapable of ruling itself; it was common sense that certain people were fit only for slavery; it was common sense that women should be denied a voice in the political process. And so, by tradition, these beliefs and practices were perpetuated.

During the eighteenth century, a new idea emerged: societies should be based on logic and reason, not tradition and common sense. The American Revolution began against this backdrop; indeed, the Declaration of Independence reads like a mathematical treatise. "We hold these truths to be self-evident" begins a list of explicit axioms, such as "All men are created equal" and "All men have the right to life, liberty, and the pursuit of happiness." From these, the Founders argued that rebellion against Great Britain was a necessary logical consequence of the actions of George III.

Unfortunately, the Founders chose to model the government of the new country on the government of the old country: a victory of common sense and tradition over reason and logic. When the Articles of Confederation proved incapable of governing the new country, a Constitutional Convention met and produced our present Constitution—another form of government based on common sense and tradition.

Fortunately, the Framers of the Constitution realized that they did not have all answers to all questions for all time. Consequently, although the Constitution requires the existence of certain institutions, such as the census, the electoral college, and jury trials, it left the details to the legislative branch of the government, with oversight by the executive and judicial branches.

The problem is that Congress operates by passing laws, leaving it to the president and the U.S. Supreme Court, and ultimately the court of public opinion, to decide whether the law is a good one. As a result, badly conceived laws persist, and unless they are reviewed and overturned, they continue to cause harm to all who live under them. Every day a bad law remains in force, it undermines faith in the rule of law—with potentially catastrophic results. Every effort must be made, then, to assess legislation before it is implemented and to continuously assess it while it remains in force.

This can be done in two ways. First, we might rely on tradition to assess legislation before it is passed and on common sense to evaluate it. Tradition says that what worked in the past will work today; thus what worked in mostly agrarian, mostly rural eighteenth-century America should work in mostly industrial, mostly urban twenty-first-century America; what worked in the age of flintlocks and sabers will work in the age of machine guns and sarin; what worked when most of a population of three million were white persons of European descent will work when most of a population of three hundred million are not. As for a law's effectiveness, it is common sense that the census must be an exact to-the-person count; that the electoral college distorts the popular vote; and that draconian punishments deter crime.

The other approach is through the reason and logic espoused by Enlightenment philosophers like the Marquis de Condorcet. In his *Essay on the Application of Probability to Decisions Made by a Plurality of Votes* (1785), Condorcet gave expression to a dream: Mathematics can tell us how to build a better society.

First, mathematics allows us to predict what would happen if we changed the very foundations of society, to answer the "What if" question that is the origin of all reform. What if we used proper statistical methods to conduct a census instead of trying to obtain a to-the-person exact count? What if we abandon the electoral college and substitute a direct popular vote? What if we replace the 12-person jury with a 3-person tribunal?

Next, we can continuously assess the effectiveness of our societal institutions in an objective fashion. For decades, the debate over capital punishment focused on whether it was morally justified; whether the United States should remain in the same club as North Korea and Iran; whether society should continue to feed, house, and clothe convicted murderers. And for decades, capital punishment persisted. It was not until the opponents of capital punishment enlisted the help of mathematics that they won a national moratorium from 1972 to 1976; and even though capital punishment returned in 1976, the philosophers, aided by mathematics, have managed to repeal capital punishment in eight states.

In the following chapters,[1] we shall see what mathematics has to say about the institutions required by the Constitution but designed by lawmakers, and the limitations on government embodied in the Bill of Rights. In some cases, mathematics supports the paths suggested by tradition and common sense. In others, it shows us a better way. In all cases, we ignore the findings of mathematics at our peril.

1. Note: The book's chapters are numbered based on the article and section of the U.S. Constitution, so, for example, chapter 2.1 is about Article 2, Section 1 (concerning the election of the president), and chapter A4.1 is about Amendment 4 (concerning the freedom from unreasonable search and seizure).

Part I / The Articles of the Constitution

Stand Up and Be Estimated

Representatives and direct Taxes shall be apportioned among the several States which may be included within this Union, according to their respective Numbers, which shall be determined by adding to the whole Number of free Persons, including those bound to Service for a Term of Years, and excluding Indians not taxed, three fifths of all other Persons. The actual Enumeration shall be made within three Years after the first Meeting of the Congress of the United States, and within every subsequent Term of ten Years, in such Manner as they shall by Law direct. The Number of Representatives shall not exceed one for every thirty Thousand, but each State shall have at Least one Representative.

—U.S. CONSTITUTION, ARTICLE 1, SECTION 2

In 2000, the U.S. Census Bureau spent $6.5 billion and employed 900,000 people to count the population of the United States. They found the population of the United States, for apportionment purposes, to be 281,424,177 persons, of whom 2,236,714 lived in Utah and 8,067,673 in North Carolina. Based on these figures, Utah received 3 congressional representatives and North Carolina 13. However, if Utah had just 856 more residents, or North Carolina had 3,086 fewer, Utah would have received 4 representatives and North Carolina 12.

Utah launched two lawsuits to try to wrest a seat from North Carolina. The first lawsuit was based on the purely political question of who should be counted. The Census Bureau includes federal employees and members of the military posted overseas as part of the population of their home state; as a result, North Carolina received an additional 18,360 residents, while Utah received 3,545. Utah argued that the Census Bureau should also count Mormon missionaries living overseas as residents. The U.S. District Court rejected this argument on November 5, 2001, and the U.S. Supreme Court declined to hear an appeal.

Probability and Statistics

Utah's second lawsuit centered around the mathematical question of how the population should be counted. To understand the problem, consider a much simpler task: Find the total resident population of an apartment complex like Parkchester in the Bronx, New York, with more than 12,000 units spread out over 171 buildings and 129 acres. Suppose we had a similar but simpler complex, with 10,000 one-bedroom units. How could we determine its total resident population?

One approach is to visit every apartment, knock on every door, and count the number of occupants. But this is time-consuming, which means that it will be expensive. Moreover, it's unlikely to yield an accurate count, because people move, both between apartments and in and out of the complex. A person might be counted twice, by two different census takers working at different times; or might not be counted at all, if he or she happens to be out of the apartment when the census takers arrive.

As an alternative, we can rely on probability and statistics. A *random experiment* is one whose outcome is in practice unpredictable. For example, flipping a coin and seeing whether it lands heads or tails is a random experiment whose outcome is "heads" or "tails."

It's important to understand that *random* does not mean *reasonless*. Whether a coin lands heads or tails is completely determined by well-understood laws of physics: the coin moves in a predictable fashion under the influence of forces like gravity and air resistance. We could imagine carefully measuring the exact dimensions of the coin and how it is placed in the hand, the air temperature and wind speed, the strength with which it is flipped, and so on, then computing how it will land. But in practice, it is not possible to measure these values accurately enough to predict how the coin will land before it actually lands; hence the result will be unpredictable. It is the unpredictability-in-practice that makes the experiment random.

Likewise, personal choices determine how many people live in any given apartment, but since we cannot know how many people live there until we actually count them, this too is a random experiment with outcomes ranging from 0 (an unoccupied apartment) on upward.

Intuitively, we would expect to see some numbers of occupants more often than others. Our one-bedroom apartment probably contains 0 or 1 person, possibly 2 or 3 persons, but 4 or more is unlikely. Mathematically we say that

Table 1.	Frequency Distribution for Residents of a One-Bedroom Apartment				
Occupants	0	1	2	3	4 or more
Frequency	5%	10%	60%	25%	0%

the outcome (in this case, the number of persons who actually occupy the apartment) is produced by a *random variable*, and we speak of the *probability distribution*, which assigns each possible value of the random variable a *probability* between 0 and 1. The basic question of probability can be phrased as follows: Given the true state of the world, predict what happens when we perform a random experiment.

In this case, the true state of the world corresponds to the actual number of residents in all 10,000 units. Suppose we knew these numbers: that apartment one has 0 occupants; apartment two has 1 occupant; apartment three has 2 occupants; apartment four has 2 occupants; apartment five has 1 occupant, and so on. We can put these values into a list: 0, 1, 2, 2, 1, etc. This list of values is known as our *population*.

Suppose our values give us the frequency distribution shown in table 1. How can we use this information to obtain a probability? The most common interpretation of probability is that it measures how often a particular outcome occurs when the random experiment is repeated a very large number of times. For example, if we were to knock on every door in the complex, we would find that 5% of the time, we'd observe an unoccupied apartment; 10% of the time, we'd observe an apartment with one occupant; 60% of the time, we'd observe an apartment with two occupants; and so on. It follows that the *probability* that an apartment has 0 occupants should be 5%; the probability that an apartment has 1 occupant should be 10%; and so on.

Sampling the Population

Probability begins with some knowledge or assumptions about the true state of the world and predicts what observations we will make: in this case, we begin with the frequency distribution for the number of occupants in the apartments and predict what we will observe when we knock on an apartment door. But the problem is that we don't actually know the true state of the world; indeed, this is what we're trying to find. Statistics allows us to infer the true state of the world from some observations.

To see how this works, suppose we knock on a few doors and find the number of occupants. This gives us a *sample*, and our goal is to use the sample to say something about the population. For example, suppose we visited 10 apartments and wrote down the number of occupants in each. As with the population, we can view these numbers as a list, say:

0, 1, 2, 2, 1, 2, 2, 2, 2, 3

We observe that, of the apartments visited, $1/10 = 10\%$ are empty, $2/10 = 20\%$ have a single occupant, $6/10 = 60\%$ have two occupants, and $1/10 = 10\%$ have three occupants. Based on this sample, we might conclude that, of the 10,000 units in the apartment complex, 10% (1,000) are vacant; 20% (2,000) have one occupant; 60% (6,000) have two occupants; and 10% (1,000) have three occupants, for a total of 17,000 residents. We would hesitate to claim that there are *exactly* 17,000 residents, so we would probably say that we "estimate" there are "approximately" 17,000 residents. Mathematical statistics allows us to refine this statement.

To begin with, it's helpful to consider a different way to obtain the figure of 17,000 residents. Since our sample of 10 apartments had a total of 17 residents, we might say that the *mean* number of residents is $17/10 = 1.7$. The mean is commonly referred to as the *average*, but this term could refer to one of several different measures, so statisticians prefer not to use it.

The significance of the mean is most easily understood in terms of quantities that can take on any value, such as length, weight, or volume. For example, imagine that we have a set of cereal boxes, each of which contains different amounts (by weight) of cereal. The mean weight is the amount of cereal that each box would contain if we redistributed the cereal so that each box had the same amount.

Obviously, we can't "redistribute" the tenants of the 10 apartments so that each apartment has 1.7 tenants. But we can look at it from the opposite direction: If it were possible to begin with 1.7 tenants per apartment, we could redistribute these fractional persons so that each apartment had a whole number of persons. Because we are redistributing but not adding, the total number of residents will be unchanged.

Our extrapolation to the population can then be expressed as follows: Based on our *sample mean* of 1.7 residents per unit, and the fact that there are 10,000 units, we predict that there are $10,000 \times 1.7 = 17,000$ residents.

Now let's consider the problem from the probability point of view. Remember that table 1 represents the true state of the world. It follows that 5% (5,000) of the apartments are empty; 10% (1,000) have one occupant; 60% (6,000) have two; and 25% (2,500) have three. This means there will be a total of 20,500 residents in 10,000 units. The *population mean* will then be 20,500/10,000 = 2.05 residents/unit. Note that there is a difference between our population mean (2.05) and our sample mean (1.7).

To do anything more than claim that the sample mean is "about" equal to the population mean, we must introduce another concept: the *standard deviation*. This is a measure of how far the data values are from the mean (population or sample) and is calculated using a simple formula. In this case, our population standard deviation is about 0.7399. We can also compute the sample standard deviation: 0.8233.

Consider the sample mean. Its actual value will depend on the number of occupants of the apartments we visit, and since these are produced by a random variable, it follows that the value of the sample mean will be unpredictable-in-practice: in other words, it too can be described using a random variable with some probability distribution.

In general, the probability distribution of the sample mean will depend on the probability distribution of the underlying random variable. However, the *Central Limit Theorem* tells us that a very remarkable thing happens with the sample mean: as the size of our sample increases, the probability distribution for the sample mean more and more closely resembles the *normal probability distribution*, with mean equal to the population mean and standard deviation equal to the *standard deviation of the sample mean*, which we will designate SD. The SD, which should not be confused with the sample standard deviation, is based on the population standard deviation and the size of the sample: for our sample of 10 apartments, the SD will be 0.2339.

The fact that the probability distribution for the sample mean closely resembles the normal distribution, informally referred to as the "bell curve," allows us to make the following claims:

1. Around 68% of the time, the population mean and the sample mean will be within one SD of each other.
2. Around 95% of the time, the population mean and the sample mean will be within two SDs of each other.

Table 2.   Maximum Reasonable Error for Different Sample Sizes

Sample Size	4,000	5,000	6,000	7,000	8,000	9,000
Error (persons)	181	148	121	97	74	49
% Error	0.88	0.72	0.59	0.47	0.36	0.24

3.   Around 99.7% of the time, the population mean and the sample mean will be within three SDs of each other.

This allows us to state a *confidence interval for the mean.* For example, since 95% of the time, the population mean is within 2 SD of the sample mean, we can give the *95% confidence interval* as the sample mean, plus or minus 2 SD: $1.7 \pm 2 \times 0.2339 = 1.7 \pm 0.4680$.[1] The value 1.7 is called the *point estimate,* while the 0.4680 is referred to as the *margin of error* and 95% is the *level of confidence.* An honest statistician will give all three, though the use of the 95% level of confidence has become so standardized that it is often omitted, much to the dismay of those who seek to apply further analysis to reported statistical data.

The fact that 1.7 ± 0.4680 is the 95% confidence interval for the population mean is often misinterpreted as meaning that there is a 95% chance that the population mean is in this interval. However, the confidence interval is something more subtle: *If* the population mean is 1.7, *then* there is a 95% probability that we will observe a sample mean in the interval 1.7 ± 0.4680. Because of this, it's more correct to interpret the confidence interval 1.7 ± 0.4680 as follows: The maximum reasonable error to expect when using 1.7 as the population mean is 0.4680 occupants per residence.

We might be dismayed by the fact that the error is so large: with 10,000 units, an error of 0.4680 occupants per residence translates into nearly 5,000 persons. But remember, our conclusions were based on a sample of just 10 units; we should instead be impressed that the error is so small. If we had visited 100 units, our SD would be 0.07363, so our confidence interval for the residential population would have a maximum reasonable error of about 1,472 persons. As the actual residential population is 20,500, this error amounts to about 7% of the population. Table 2 shows how the error decreases as the size of the sample increases.

---

1. Strictly speaking, this gives us a 95.45% confidence interval. The (statistical) 95% confidence interval uses 1.96 SD. However, as we shall see, there is almost no difference worth mentioning between the 95% confidence interval and the 95.45% confidence interval.

The Undercounting Problem

Modern mathematical statistics emerged from the work of Karl Pearson (1857–1936) and Ronald Fisher (1890–1962) at the dawn of the twentieth century. Until the mid-twentieth century, the Census Bureau could use only the most primitive statistical methods, and conducted the census by obtaining occupancy information from every known residential address, usually by sending a census taker. There are some obvious limitations with this approach. First, the residents might be absent when the census taker arrives. Second, the Census Bureau might not have all residential addresses. Consequently, the census invariably produces undercounts.

To determine the magnitude of the undercount, the Census Bureau uses a method known as *demographic analysis*. This essentially uses independent means to measure the population, then compares the estimate to the census figures: for example, information from birth and death certificates can be used to determine the population growth. The 1940 census, which showed the population of the United States to be 131,669,275, was believed to have missed about seven million people—about 5.4% of those actually counted.

Shortly after the release of the census data, World War II began, and draft registration provided an unprecedented measure of the undercount. Particularly disturbing was the fact that, while the census reported 12,865,518 blacks in the country, their Selective Service registrations suggested the actual black population was about 1.1 million higher: this was fully 8.4% of those actually counted and showed an undercount rate substantially higher than that of whites.

The differential undercount is disconcerting, since it implies that certain areas would be more strongly affected by the undercount than others. Intuition suggests that undercounting blacks would most strongly affect states with large black populations, the South in particular. As a result, the Census Bureau attempted to reduce the undercount in general and the differential undercount in particular, with a focus on the southern states. They reduced the black undercount to about 7.5% for the 1950 census, and even further down, to 6.6%, for the 1960 census. These efforts proved more successful at reducing the white undercount rate, which fell to 2.7% in 1960, ironically widening the undercounting gap between blacks and whites recorded by the census.

A Sample of Sampling

In 1970, the Census Bureau inaugurated a new procedure for conducting the first phase of the census: a massive mailing to as many residential addresses as feasible. The occupants were to fill out the forms and mail them back ("self-enumeration," a feature that first appeared in the 1960 census). Because federal money and congressional apportionment depend on the census figures, those who fail to fill out and return the census forms act against their own best interests; it is almost incidental that failure to fill out and return the forms is also a violation of federal law. Nevertheless, only 78% of the forms mailed out were mailed back.

There are several reasons a census form might not be returned. First, the residence might actually be vacant. Second, the occupants might not have received the form, or might simply ignore it. Such nonresponding residences contribute to the undercount.

In an effort to reduce the undercount, the Census Bureau implemented a second program, known as the National Vacancy Check, and sent census takers to 13,546 of the roughly 14 million nonresponding residences. Based on this sample, the Census Bureau reclassified about 8.5% of them as occupied, and added 1,068,882 persons to the population total; of these, 348,913 were in southern states.

It might seem that investigating less than 0.1% of the nonresponding residences (in the case of the National Vacancy Check) would surely produce sizable errors. But one of the more surprising results of mathematical statistics is that the relative error depends primarily on the absolute size of the sample and the underlying probability distribution, not on how large the sample is relative to the population. In particular, there is no need for the sample to be a "sizable fraction" of the population in order to obtain good results; it is only necessary that the sample be sufficiently large.

For example, a sample of 200 units drawn from our hypothetical 10,000-unit apartment complex would allow us to determine the total population to within about 5%. If instead we were considering a supersized apartment complex that was 1,000 times as large, yet with the same probability distribution as our original complex, a sample of 200 units would *still* allow us to determine the total population to within 5%.

The 5% of the residents of the 10,000-unit complex amounts to about 1,000 persons, while 5% of the residents of the supersized 10,000,000-unit apartment complex constitutes about 1,000,000 persons. If we wanted to maintain the absolute error at around 1,000 persons, we would have to increase the sample size. Here the mathematics works against us: to find the population to within 1,000 persons, we'd need to use a sample of 9,500,000 of the units!

It is worth emphasizing that the margin of error present when using sampling *only* applies to the population actually sampled. In 1970, about 78% of the households returned their census forms, providing the Census Bureau with demographic information about this group. Consequently, the error from the sampling of the National Vacancy Check would only apply to the 22% of nonresponding units. Since the Census Bureau actually recorded a population of about one million persons for this group, even a 10% error in this group's population (far larger than would be reasonable to assume, given the size of the sample) would result in an error in the *national* population of just under 0.05%.

This still amounts to about a hundred thousand persons. We might decide that this is still too large an error and insist on a larger sample of the nonresponding units to reduce the undercount. In fact, in 1976, the U.S. Congress amended the Census Act to read: "Except for the determination of population for purposes of apportionment of Representatives in Congress among the several States, the Secretary shall, if he considers it feasible, authorize the use of the statistical method known as 'sampling' in carrying out the provisions of this title."

This amendment to the Census Act is problematic for a number of reasons. First, it is ambiguous: the amendment, as written, does not *forbid* the use of sampling, but instead authorizes its use for most of the census. More problematic is that it's unclear what "the statistical method known as 'sampling'" actually refers to. *All* statistical methods use sampling in one form or another.

If we assume Congress meant to forbid the approach used by the 1970 National Vacancy Check, it follows that the Census Bureau must visit every nonrespondent household until it can obtain information about the number of residents from an occupant. Moreover, census takers must continue to visit the address until they interview an occupant or establish that the dwelling is in fact unoccupied. By 1990, census takers visited nonrespondent households up to six times before giving up. As personal visits are the most expensive way to gather data, census costs skyrocketed. Adjusted for inflation to the year 2000, the 1970 census, which used sampling and statistical methods, cost

$12 per housing unit. The 1980 census, which did not, cost $24 per housing unit; and the 1990 census, which also precluded sampling and statistical methods, cost $31 per housing unit.

To Correct . . . or Correct

The 1970 census included a second method of improving the accuracy of the census: the Post Enumeration Post Office Check (PEPOC). This program involved comparing Census Bureau residential addresses to mail delivery addresses, supplied by the U.S. Post Office, under the assumption that a person received mail at their residence. Again, visits to a sample of these addresses added 589,517 persons to the population; however, for a variety of reasons, the Census Bureau only applied the PEPOC to 16 southern states, where residential data were believed to be especially poor.

As its name indicates, the PEPOC occurred after the census data collection period ended. In effect, the Census Bureau obtained an initial population figure, but on the basis of additional data produced a more accurate population figure.

No census figure will ever be completely accurate, for no other reason than that the population is constantly changing. The important question, then, is not "Can we make the census accurate?" but rather "How much time, effort, and money are we willing to expend to increase the accuracy of the census?"

The paradox is that in order to determine how accurate the census is, we must determine the population—which we can't do without an accurate census. To resolve this, we need some independent means of determining the population. The plans for the 1980 census included a Post-Enumeration Program (PEP) to determine the fraction of the population actually counted by the census. PEP included a number of parts, the most important being dual-system estimation (DSE), the application to the census of a well-established method in wild animal studies known as *capture-recapture*.

Suppose you want to determine the number of fish in a lake. One possibility is to drain the lake and count the fish remaining. Of course, this destroys the lake and kills all the fish, so a better method is obviously preferable. The standard strategy is to catch and tag some number of fish; release them back into the lake; then at some later point catch some fish and see what fraction of the fish caught are tagged. For example, suppose we initially catch and tag 100 fish. Some time later, we catch 100 fish (the number we tag and the number we catch later need not be the same) and find that 5 of the fish we caught are tagged. A simple inference is that since 5/100 = 5% of the fish we caught are tagged, 5% of the total fish are tagged. Since we know the number of fish we actually tagged, we can use this to estimate the total population of the lake: in this case, 2,000 fish. Of course, no one would ever claim there are *exactly* 2,000 fish in the lake, and as in our apartment complex example, we can use standard statistical methods to find a confidence interval for the total fish population.

Using the same approach, the Census Bureau planned to determine the number of people missed by the census. First, the census forms are processed; this corresponds to the initial capture and tagging of the fish. Next, the Census Bureau would select two samples, of 84,000 households apiece, and determine the fraction of those households that filled out census forms; this corresponds to the second capture and comparison of the tagged fish to the untagged fish.

PEP could be used to correct the census figures, but the secretary of commerce was not legally obligated to use it. Since the differential undercount means that areas with large non-white populations would benefit from a corrected count, the mayors of Detroit and New York sued to compel the secretary of commerce to use the adjusted figures. Their lawsuits were eventually dismissed, and the 1980 census was not adjusted.

To some extent, the dismissal was legitimate: the Census Bureau had not itself reached consensus on how the population undercount should be adjusted. However, recognizing that the undercount was an important issue, the Census Bureau drew up plans to include a Post Enumeration Survey (PES) that *would* be used to correct the 1990 census. Unfortunately, as the census date approached, there was no consensus on how to implement the plan.

Exactly what happened afterward is subject to some controversy, but the purely objective facts are these. The director of the Census Bureau, Jack Keane, proposed using the PES to adjust the census figures. But in late 1987, Secretary of Commerce Calvin W. Verrity overrode Keane's decision. Barbara Bailar and Kirk Wolter, two senior statisticians and proponents of the use of PES to adjust the population figures, resigned.

Administratively, both the director and the secretary of commerce are political appointees, while the census department staff itself consists of apolitical civil servants. This led to charges and countercharges of political interference with a scientific enterprise and—no surprise—lawsuits over whether the

1990 census figures should be adjusted. In July 1989, the new secretary of commerce (Robert Mosbacher, appointed by President George H. W. Bush) agreed to allow the Census Bureau to produce a set of adjusted figures.

The preliminary figures were released in December 1990, as required by law. Determining the adjusted figures took more time than anticipated, but by June 1991, the figures were ready. In the opinion of Barbara Bryant, the new director of the Census (Keane resigned in 1988), as well as seven of the nine census officials on the Undercount Research Steering Committee, the adjusted figures were more accurate than the preliminary figures and would have increased the U.S. population by about 5.3 million. Nevertheless, the secretary of commerce rejected the corrected figures. This led to a new round of lawsuits, but in 1996, the Supreme Court, in a unanimous decision, held that the secretary of commerce was not required to use the corrected counts.

What prompted Mosbacher to dismiss the attempts of the Census Bureau to provide a more accurate count of the national population, and to override the professional judgment of trained statisticians and demographers? Mosbacher gave a number of reasons, but among them was speculation that allowing counts corrected by accepted statistical methods would "open the door" to allowing counts to be altered for political gain. Indeed, Representative Newt Gingrich of Georgia signed a letter urging the Census Bureau to use the adjusted figures—if they did so, Georgia would gain a congressional seat. Meanwhile, Representative Christopher Shays—who, like Gingrich, was a Republican—opposed adjustment, because his home state of Connecticut would lose a seat.

Gingrich and Shays, at least, had a rational basis for their stance on whether or not to use the adjusted figure: although it was politics as usual, at least it was politics based on real data. In contrast, Jim Nicholson, chairman of the Republican National Committee, claimed in 1997 that census officials would add four and a half million *Democrats* because of adjustment of census data through the use of a "complex mathematical formula."

However, if Nicholson's claim had any merit (and there is no evidence that it did), then either the four and a half million persons *already* existed in the population and weren't being counted, or the "complex mathematical formula" was badly flawed. In the first case, denying the Census Bureau the authority to adjust the figures is tantamount to requiring them to report inaccurate figures. In the second case, mathematics, unlike politics, always takes place in the light: If the Census Bureau was using a flawed formula, it would either have to present the formula for all to see, at which point its flaws would become readily apparent; or it would have to conceal its formula from prying eyes, which would itself draw scrutiny.

Imputation: Hot and Cold

Even if money is no object, and the Census Bureau could afford to have census takers visit nonresponding households a thousand times if necessary, the problem of missing data would remain: some residences will never be occupied when a census taker arrives. Since the residence has some number of occupants, we must *impute* a value—possibly zero, if the residence is in fact unoccupied—for this number.

The obvious solution is to impute the mean number of residents found from our sample. In fact, this is exactly what we did in our earlier discussion: the mean number of residents in our sample of 10 apartment units was 1.7 occupants, leading to an estimated total residential population of 17,000 based on imputing 1.7 residents per remaining unit.

There are relatively minor issues associated with imputation. First, the imputed value should be based on similar units: if our apartment complex included two-bedroom units, we would hesitate to impute the value of 1.7 for the number of occupants of these units. However, this just means we would need to obtain a sample of two-bedroom numbers to find a value to impute for them.

A more important question is whether to use *hot deck imputation*, which is based on the data currently being gathered, or *cold deck imputation*, based on data gathered from previous censuses. The advantage to cold deck imputation is that (in theory) any errors, such as miscounts or overcounts, have been removed. The disadvantage is that in the 10 years since the last census, the demographic characteristics of a region may have undergone dramatic changes. For this reason, the census bureau has traditionally favored hot deck imputation.

The 1980 census began, as in 1970, with a massive mailing of census forms to about 96% of the residences in the United States. Roughly 80% of these forms were returned. Census workers went to nonresponding households to try to obtain information about the number of residents, and in most cases succeeded. However, there remained a small group of residences from which no information could be obtained.

For example, suppose there are three houses along a road. One is vacant, and confirmed to be vacant. A second is occupied, with a family of four. If, after repeated attempts, no information could be obtained from the third unit, the Census Bureau would have to impute a value for the number of occupants. In 1980, they used the following strategy. First, if there was evidence of occupation, then the number of residents was set equal to the number of residents of the most recently processed occupied unit. In this case, since the most recently processed occupied unit had four residents, a value of four would be imputed for the number of residents of the nonresponding unit.

However, if the occupancy status was unknown, both the occupancy status and number of residents would be determined from the most recently processed unit. In this case, if the nonresponding unit was visited after the house of the family of four, the imputed status would be "occupied" and the imputed number of residents would be four. On the other hand, if the nonresponding unit was visited after the unoccupied unit, its imputed status would be "vacant" and the imputed number of residents would be zero.

Since every attempt was made to obtain data from nonresponding units, the total population imputed was very small: about 0.3%. However, this is enough to alter a state's congressional delegation, and because of imputation, Indiana lost one seat in the House of Representatives and Florida gained one. Thus the state of Indiana challenged the use of imputation in *Orr v. Baldrige* (1985).

The case would be heard by the U.S. District Court for the Southern District of Indiana, which made two important rulings. First, it ruled that the Census Act did in fact prohibit the use of sampling. But it also held that, barring sampling, the Census Bureau had the discretion to use hot deck imputation as a means of determining the population: this decision was a purely administrative one, well within the authority granted to the Census Bureau by Congress and thus not subject to judicial review.

Sampling Again

Following the 1990 census, Congress passed the Decennial Census Improvement Act of 1991, instructing the secretary of commerce to call upon the National Academy of Science for ways to make the census more accurate. Among other things, the Academy was specifically directed to consider the appropriateness of sampling methods alongside traditional methods. Based on the National Academy's recommendations, the Census Bureau drew up a plan for the 2000 census.

First, as it had since 1970, the Bureau planned to send out questionnaires to all known residential addresses. Based on previous results, the Census Bureau could expect to obtain data from about 70% of the residential addresses. But rather than plan visits to every one of the nonresponding units, the Census Bureau would select enough of them to ensure that information could be obtained from 90% of the housing units in each census tract (an administrative unit of about 4,000 persons). For example, if a census tract had 1,000 units but only a 50% response rate, so that census forms had been received from only 500 of its 1,000 residences, workers would be sent to 400 of the remaining residences to obtain 90% coverage. In contrast, a census tract that had an 80% response rate would only have an additional 100 residences visited. The data from these residences would then be used to infer information about the remaining residences.

Next, the bureau would employ an Integrated Coverage Measure, surveying about 750,000 residences in person. As with 1990's PES, a capture/recapture method would be used to infer data about persons who did not return census forms, but this time the numbers would be used to correct the census data and provide the most accurate return possible.

Congress objected, and in 1997 passed legislation that forbade the Census Bureau from using sampling or *any* statistical technique to determine the population for apportionment. This would have made the Census Bureau's task impossible, so it was successfully vetoed by President Bill Clinton. However, this left the ambiguous provisions of the 1976 Census Act amendment in place, and the House of Representatives filed suit against the Department of Commerce to prevent it from using statistical sampling in the 2000 census.

The Supreme Court heard *House of Representatives v. Department of Commerce* in 1998, and issued its ruling on January 25, 1999. Although it agreed that the wording of the amendment was ambiguous, it held that past censuses had not used statistical sampling but instead required census takers to attempt to visit every household. As a result, the Census Bureau was forced to make an effort to visit, in person, every nonresponding household, and had to petition Congress for an extra $1.7 billion to cover the new costs. The 2000 Census ended up costing $61 per housing unit, for a total cost of nearly $7 billion.

Imputation and Sampling

Regardless of how much time, effort, and money the Census Bureau spends to obtain information from nonresponding households, there are *still* units for which no information could be obtained. As in previous censuses, it used hot deck imputation to infer the missing information.

It might seem that *Orr v. Baldrige* settled the question of whether or not the Census Bureau could use hot deck imputation. However, *House of Representatives v. Department of Commerce* changed the nature of the game by specifically forbidding sampling. In *Utah v. Evans* (2002), the state of Utah challenged the mathematical methods used by the Census Bureau on two grounds. First, it pointed to the clause in the Constitution referring to an *actual* enumeration, arguing that this wording meant that census takers were required to visit each and every residence in the country. Second, it argued that hot deck imputation was a form of sampling. If it prevailed in either argument, hot deck imputation would be deemed unconstitutional.

The first argument hinges on a point of grammar: is enumeration a verb or a noun? If it is a noun, then "actual enumeration" simply requires that the census be conducted within three years of the meeting of the First Congress. But if it is a verb, then "actual enumeration" specifies how the census is to be conducted. The Supreme Court took the former viewpoint: the Constitution does not require a person-by-person count, and Congress is free to specify how the census should be conducted.

Next, the Supreme Court ruled that imputation is not sampling. This conclusion was based on two factors. First, sampling *begins* with no intention of collecting data from the entire population. Next, the units from which data are collected are chosen using statistically valid sample-selection methods. Thus, since the residences whose occupancy status was imputed were not chosen by the Census Bureau in any fashion, imputation cannot be sampling. As a result, hot deck imputation was once again affirmed and statistical sampling was once again rejected.

The silver lining is that the Supreme Court rejected sampling on statutory grounds, not constitutional grounds. This leaves the actual methods used by the Census Bureau to the discretion of Congress. Currently, sampling is forbidden, because Congress forbade it. But census costs continue to escalate: the 2010 Census cost $13 billion, or about $100 per housing unit (equivalent to $79 in terms of year 2000 dollars). Faced with the ever-rising cost of conducting the census, Congress could just as easily decide to allow sampling.

The problem, as lawsuits against the Census Bureau show, is that slight disagreements over the population of a state can cause it to lose a congressional seat—which guarantees that at least one member of Congress will not be re-elected. As a result, there are endless debates over the best way to count the population. However, as we will see in the next chapter, the problem has very little to do with how the population is counted, and everything to do with how the number of representatives is determined.

(Nearly) Equal Representation

Congress's failure, then, to make the inequality [between congressional district sizes] slightly less is within its discretion to balance many factors . . . that cannot then be reviewed by elementary arithmetic.
—JUDGE LESLIE H. SOUTHWICK, FIFTH CIRCUIT COURT OF APPEALS, *CLEMONS V. DEPARTMENT OF COMMERCE* (2010)

In 1990, the state of Montana sent two representatives to Congress, while the state of Washington sent eight. But in the reapportionment following the census, Montana lost a seat and Washington gained one. Montana sued, arguing that the method used to apportion seats violated the Constitution. The U.S. Supreme Court heard *United States Department of Commerce v. Montana* on March 4, 1992.

Montana's claim was based on the following facts. Under the new apportionment, Montana's single congressman represented 803,655 persons. On the other hand, each of Washington State's nine representatives oversaw a constituency of 543,105 persons, meaning the Montana congressman represented 260,550 more people. But if a representative were transferred from Washington to Montana (incidentally restoring the 1990 apportionment), Montana's two congressmen and Washington's eight would represent 401,828 and 610,993 persons respectively, a difference of 209,165. Since the pre-1990 apportionment produced a smaller difference, Montana argued that the post-1990 apportionment violated the implied requirement that congressional districts be as nearly equal in size as possible.

Simply put, the *apportionment problem* is as follows: Each state is to have a certain number of representatives in Congress, based on the state's population. But how should that number be determined? The importance of this question can be measured by the fact that it was the basis of the first proposed amendment to the Constitution, and it would result in the first presidential veto in American history.

The Hamilton-Jefferson Debate

The Constitution gives very little guidance on the apportionment problem. In fact, there are only two restrictions. First, the number of representatives may not exceed "one for every thirty thousand" persons. Population growth has made this figure meaningless, with the number of representatives less than one per five hundred thousand persons.

The second restriction is more important: the number of representatives is to be allocated based on population. However, while the Constitution specifies how the apportionment population is to be computed from the actual population, with free persons, slaves, and Indians counted as 1, 3/5, and 0 persons respectively, it does not indicate how these apportionment populations translate into representatives. The first proposed amendment to the Constitution outlined how representatives would be allocated to the states, but as of 1791, when Congress had to apportion congressmen to the states, the amendment had not been ratified by the necessary number of states.

Since the amendment had been approved by Congress in the first place, they went ahead and used its basic structure, which might be described as follows. First, choose an ideal district size N; the actual amendment specified $N = 30{,}000$. Next, divide each state's population by the ideal district size to obtain that state's *quota*. Since this will in general produce a whole number plus some fractional amount, and since we cannot send a fraction of a person to Congress, this number must be rounded in some fashion to obtain the state's congressional apportionment. Thus the apportionment question can be solved by answering two questions: How big should a congressional district be, and what should we do with the fractional part of the state quotas?

According to the 1790 census, the United States had an apportionment population of 3,615,920. If we divide this by 30,000, the number suggested by the Constitution, we obtain approximately 120.531. Under the ordinary rules of mathematical rounding, this should be rounded up to 121. However, if there were 121 representatives, there would be more than one representative for every 30,000 persons, which would violate the constitution. Thus we should round down and allocate at most 120 representatives to the various states.

Next, we can divide each state's apportionment population by 30,000 to obtain the quotas in table 3, where we list the states in order based on the *fractional part* of their quota: thus New Jersey, with quota 5.98567 and fractional part 0.98567, is ranked higher than Connecticut, with quota 7.89470 and fractional part 0.89470. As before, if we round any of these quotas up, we will have more than one representative for 30,000 people, so we should round each of these quotas down. But this results in the allocation of only 112 seats: 8 seats remain unallocated.

We could simply ignore the disparity, and leave the eight seats unassigned. However, this means that some states lose more than others: for example, rounding New Jersey's quota of 5.98567 down to 5 was equivalent to losing nearly an entire seat. In contrast, Virginia lost only 0.01867 seats by rounding down.

To remedy this situation, Alexander Hamilton, then secretary of the treasury, suggested that the eight seats needed to bring the total number of representatives to 120 should be given to the eight states with the largest fractional part; these are the first eight states on the list in table 3; coincidentally, these

Table 3.   Population and Apportionment for $N = 30,000$

State	Representative Population	Quota for $N = 30,000$	Round Down Apportionment
New Jersey	179,570	5.98567	5
Connecticut	236,841	7.89470	7
South Carolina	206,236	6.87453	6
Delaware	55,540	1.85133	1
Vermont	85,533	2.85110	2
Massachusetts	475,327	15.84423	15
North Carolina	353,523	11.78410	11
New Hampshire	141,822	4.72740	4
Pennsylvania	432,879	14.42930	14
Georgia	70,835	2.36117	2
Kentucky	68,705	2.29017	2
Maryland	278,514	9.28380	9
Rhode Island	68,446	2.28153	2
New York	331,589	11.05297	11
Virginia	630,560	21.01867	21
Total	3,615,920		112

are the eight states that would have had their fractional parts rounded up, though this will not always occur. To his credit, Hamilton advocated this plan even though his home state of New York did *not* benefit; we may take this as a measure of his devotion to the principle of *national* government. Hamilton's plan passed both Houses, and went to President George Washington for signature.

Thomas Jefferson objected to the plan on the grounds that it was based on "a difficult and inobvious doctrine of fractions." Given that the "inobvious doctrine of fractions" involved such problems as determining whether 0.84423 (Massachusetts' fractional part) was higher than 0.42930 (Pennsylvania's fractional part), we might question Jefferson's assessment. Certainly Jefferson himself, whose personal library shows familiarity with higher mathematics, including calculus and Condorcet's work on probability, would have had no problems with the mathematical questions. Washington, a surveyor, would have been even less intimidated by mathematical computations.

Nevertheless, Washington chose to veto the original apportionment bill, for several reasons. The most important was that assigning extra seats to some states gave them *more* than one representative for every 30,000 persons. For example, the apportionment bill would have given Delaware two congressmen, each of whom would represent 27,770 persons. Consequently, Washington had doubts about the constitutionality of the proposed apportionment. It's not clear that this would have been the case, since the Constitution leaves unclear whether the 30,000-person limit applies to the total number of congressmen, or the congressmen in each state. Under some philosophies of executive authority, Washington should have approved the plan and let the Supreme Court decide its constitutionality.

Congress tried to override the presidential veto, but failed to muster the necessary two-thirds vote. It submitted a different plan, originating in the Senate and supported by Thomas Jefferson. The Jefferson plan used an ideal district size of $N = 33,000$ and rounded the state quotas down. This assigned 105 seats.[1] The next four apportionment bills were handled in the same way, with the value of N steadily increasing.

---

1. It may also be relevant that, under Hamilton's plan, Virginia—the home state of both Washington and Jefferson—sent 21 representatives to a Congress of 120: 17.5% of the total. But under Jefferson's plan, Virginia sent 19 representatives to a Congress of 105: 18.1% of the total.

Webster, Interrupted

The 1840 reapportionment was handled slightly differently, using a method proposed by Daniel Webster, senator from Massachusetts, in 1832. There were two crucial differences. One was relatively minor: rather than rounding the quotas down to obtain the number of representatives, they were rounded up or down according to standard mathematical rules.

The second dealt with the value of N, the ideal district size. On the one hand, Congress could simply decide how big each district should be, then allocate seats accordingly. As the 1790 apportionment shows, this can lead to some odd results, which we may view in terms of a difference between what we predict and what we observe. We would predict that dividing a national population of 3,615,920 into 30,000-person districts would give us 120 representatives; yet, when we actually perform the apportionment, we end up with just 112. There's nothing wrong with this, mathematically or legally but the difference between the prediction and the observation is unsettling.

This leads to the question: Which is more important, the size of the district or the number of congressmen? Washington and the other Founding Fathers held that the size of the district was more important: If the districts were too large, the representatives would be too distanced from their constituents; moreover, the smaller the number of representatives, the easier it was to influence a majority of them. Today, we might add another constraint: the larger the size of the district, the greater the expense of campaigning, which in turn increases the importance of fund-raising and makes candidates more beholden to donors.

However, the 1830 apportionment yielded a 240-member House. Complaints about not being able to hear speakers, or rowdy behavior by congressmen in the back rows, or the practical difficulty of fitting more than 200 persons into a single chamber, made the issue of House size more important than district size.

In 1840, Congress first passed a bill that reduced the size of the House to 223 members—the only time in U.S. history Congress downsized itself and, in view of the fact that this guaranteed that at least 17 of those currently in the House would not be re-elected, a remarkable political achievement. With a national population of 17,069,453, this would give each district $17,069,453 \div 223 = 76,545$ persons. However, dividing the state populations by

76,545 and rounding would result in too few representatives being assigned. This would lead to the problem of allocating the remaining seats.

One solution is to use a *modified divisor*. This is essentially Jefferson's system, with an important restriction: rather than choosing a modified divisor and a rounding rule, which lets the number of congressmen float around some value, the 1840 apportionment chose the rounding rule and the number of congressmen, then found a modified divisor that worked.

To understand this process, let's suppose we have a national population of 1,200 persons, divided into three states with respective populations 530, 380, and 290. If we want to elect 8 representatives, then the ideal district size would be $1,200 \div 8 = 150$ persons. The first state's quota be $530 \div 150 \approx 3.53$; the second state's quota would be $380 \div 150 \approx 2.533$; and the third state's quota would be $290 \div 150 \approx 1.933$. However, if we followed standard rules of rounding, the states would receive 4, 3, and 2 representatives apiece, and our Congress would have a total of $4 + 3 + 2 = 9$ members: one more than we want.

Since we have too many representatives, we choose a larger divisor. This requires some trial and error: for example, if we try $N = 160$, we obtain an apportionment of 3, 2, and 2, which apportions too few representatives. If we try $N = 155$, we again have an apportionment of 3, 2, 2, so we must try again. A third try, with $N = 152$, gives us an apportionment of 3, 3, and 2, for the desired total of 8 representatives.

As the preceding shows, finding a suitable divisor can require a lot of computations. This is easy to do—with computers and spreadsheets, neither of which existed in 1840. We have no record of how much work it took to find the modified divisor (which turned out to be 70,680), but immediately after the 1840 apportionment, Samuel F. Vinton, representative from Ohio, proposed that Hamilton's method be used for successive apportionments; Congress and the president agreed, and the method was used between 1850 and 1900.

However, doubts about Hamilton's method began to surface. Following the 1880 census, Congress considered increasing the size of the House, and directed Charles W. Seaton, the chief clerk of the Census Office, to find the number of representatives due each state for every possible House size, from 275 to 350. He did so, and discovered a remarkable fact: If there were 299 representatives, Alabama would be given 8 seats, but if there were 300 representatives,

Alabama would only receive 7 seats. The observation that a state can *lose* representation when the House size *increases* is now known as the *Alabama Paradox*.

Webster's method, and in general any divisor method, is immune to this paradox. Moreover, by 1880, mechanical computing devices had become available, which eased the problem of computing each state's quota for the different possible divisors. Seaton himself had invented a mechanical tabulator for speeding the processing of census results, and mechanical calculators were beginning to become available.[2] Congress returned to the Webster method for the 1910 reapportionment. The next year, Congress passed the Apportionment Act of 1911, which limited the size of the House to 435 members. Except for brief periods of time following the admission of Alaska and Hawaii, the number of congressmen has remained at 435, despite the fact that the U.S. population has tripled in the past century.

The Polestar of Equal Representation

Although the Webster method would be used in the 1910 apportionment, a new round of arguments began over how to compute a state's apportionment. Because the total number of congressional seats is fixed at 435, one state's gain is another state's loss, and the different apportionment methods are perceived, rightly or wrongly, as benefiting differently sized states in different ways.[3] Congress found itself embroiled in a never-ending series of debates over the choice of divisors and methods. The problem was exacerbated by an important demographic shift: World War I had caused a mass migration from the southern states to industrial centers in the north. As a result, representatives from southern states were concerned that any alteration of the apportionment could shift the balance of power in Congress. The debate proved

2. It worked, but the task was enormous and the results of the 1880 census were not certified until 1887. This led Herman Hollerith, a census employee, to invent an electromechanical device for tabulating census records, which was used for the 1890 census. Hollerith eventually left government service and founded the company that would become IBM.

3. This is the "large vs. small" controversy, the belief that the more populous states share interests and vote alike. But as James Madison pointed out during the Constitutional Convention, the size of a state was hardly a determinant of its politics. Thus (in the modern United States) Texas has more in common with New Mexico than with New York. For that matter, a representative from Dallas probably has more in common with a representative from New York City than a representative from Laredo.

so divisive that Congress failed to produce an apportionment bill following the 1920 census, in violation of its constitutional obligations.

To solve the apportionment problem, Congress requested that the National Academy of Sciences appoint a committee of experts to consider the subject. The mathematical question may be posed as follows: suppose we have two states, with populations A, B respectively. What allocation of representatives a, b is most nearly fair?

The problem is that "fair" is not a mathematical concept; it is a political one. In the words of the Supreme Court (in *Department of Commerce v. Montana*), we need to identify the "polestar of equal representation." One possibility is that the fraction of congressmen assigned to a state should equal the fraction of the national population resident in the state. For example, the 1790 census gave Virginia 17.43% of the total population, so in a perfectly fair division, it would have received 17.43% of the representatives. If (as occurred in 1790) there are to be 105 congressmen, this means that Virginia should receive 18.31 of them. A method *satisfies quota* if a state receives either its upper quota (the next higher whole number, in this case 19) or lower quota (the next lower whole number, in this case 18). Since Virginia actually received 19 representatives, Virginia's 1790 apportionment satisfied quota.

The Hamilton method always satisfies quota, but the Jefferson method does not. For example, if the 1790 apportionment had used a district size of $N = 37{,}000$, it would have assigned 88 representatives to the states, and Virginia's fair share would have been 15.34 representatives; this gives an upper quota of 16 and a lower quota of 15. But using $N = 37{,}000$ would give Virginia 17 representatives. Thus Jefferson's method can *violate quota*.

The possibility of a quota violation is sometimes presented as an objectionable feature of an apportionment method, and one method, the Balinski-Young, always satisfies quota. While we are free to choose the features we find objectionable, there is nothing sacrosanct about quota. Standard rounding rules require rounding to the nearest whole number, but there are many contexts when we fail to do this. For example, we routinely round times to the nearest 15-minute interval: we might say that it is "3:30" when the actual time is 3:26:37.

Moreover, there is legal precedent for permitting quota violations. In 1820, New York's congressional quota was 32.503, but it actually received 34 seats, while in 1830, its quota was 38.593 and it received 40 seats. Because quota has been violated in the past, trying to make it the "polestar of equal representation" is legally challenging.

There is another reason why we might not want to make quota too important. Suppose one state loses population while another state gains it. We would be very disturbed if our apportionment method *increased* the size of the congressional delegation of the state that *lost* population, while *decreasing* the size of the congressional delegation that *gained* population: this would be the *population paradox*. Yet Hamilton's method, and in fact *every* method that satisfies quota, can produce such a situation. Ironically, this fact would be proven by Balinski and Young themselves.

What can we use as our polestar of equal representation? Quota compares a state's population A and congressional delegation size a to the national population and congressional delegation size. Instead, we might compare one state's population A and congressional delegation size a to another state's population B and congressional delegation size b. This not only gives us a way to measure malapportionment, but a logical and consistent means of redressing it, as follows. Suppose we identify the existence of a malapportionment between two states. If transferring a seat from one state to another reduces the malapportionment, we should make the exchange. Moreover, if no possible transfer reduces the malapportionment, then the apportionment is the best possible.

It remains to decide on a measure of malapportionment. Suppose two states have populations A, B respectively, and currently have a, b congressmen. Then A/a and B/b are the average sizes of the congressional districts in the two states, if we assume all the districts in a state have the same population. We might want to make the difference in district sizes as small as possible. Alternatively, the fractions a/A and b/B might be viewed as the number of representatives per person, and we might want to make the difference between these fractions as small as possible. Yet again, the fractions A/B and a/b represent the ratio of state populations and the ratio of the congressional delegations, and we might want to make the difference between these ratios as small as possible.

Historically, focus on malapportionment has centered on the district size, and we might take this as our starting point. But there remains one final problem: in general, there are two ways to measure a difference between two quantities. First, we can consider the *absolute difference*: the greater quantity minus the lesser. Next, we can consider the *relative difference*: the absolute difference divided by some reference quantity.

For example, consider the 1790 apportionment, which gave Delaware 1 seat and Virginia 19. Delaware's lone congressmen represented 55,540 persons, while each of Virginia's represented $630,560 \div 19 \approx 33,187$ persons. The absolute difference is thus $55,540 - 33,187 = 22,353$ persons.

Relative difference is somewhat more complicated, because we have to decide what the difference is relative *to*. The usual choice is the smaller of the two quantities being considered. In this case, since the Virginia district of 33,187 is smaller, our relative difference would use this number as our reference quantity, giving us a relative difference of $22,353/33,187 \approx 67\%$.

Suppose we transfer a seat from Virginia to Delaware. Then the Delaware districts will have $55,540 \div 2 = 27,770$ persons, while the Virginia districts will have $630,560 \div 18 \approx 35,031$ persons. The absolute difference will be $35,031 - 27,770 = 7261$ persons. Since the Delaware district is now the smaller, we will use it as our reference to find the relative difference as $7261/27,770 \approx 26\%$. Because transferring a seat from Virginia to Delaware would have reduced both the relative and absolute differences, it should have been done.

What if we try to transfer a second seat, which would give Delaware 3 congressmen and Virginia 17? In this case, Delaware (with three congressmen) would have districts with 18,513 persons, while Virginia's districts would have 37,092 persons. This gives us an absolute difference of $37,092 - 18,513 = 18,579$ and a relative difference of $18,579/18,513 \approx 100\%$. Since this second transfer would increase the absolute difference (from 7,261 to 18,579) and the relative difference (from 26% to 100%), we should not make the transfer.

In this case, a transfer of seats either increased both the absolute and relative differences, or decreased both; and in either situation, the course of action was clear. However, it's possible for one to increase while the other decreases. The 1990 apportionment produced a Montana district size of 803,655 persons and a Washington district size of 543,105 persons. This was an absolute disparity of 260,550 persons and a relative disparity of 48%. But if we transferred a congressional seat from Washington to Montana, Montana's district size would fall to 401,838 and Washington's would rise to 610,993 persons, producing a smaller absolute difference (209,165) but a larger relative difference (52%). We must decide which is more important, but the problem is that there is no clear consensus.

For example, if we have two objects, one of which has a weight of 20 pounds and the other a weight of 22 pounds, we might judge the two objects to have about the same weight; on the other hand, if we have a 1-pound object and a 2-pound object, we would almost certainly identify the latter as "significantly" heavier than the former. In this case, the absolute differences are 2 pounds and 1 pound, while the relative differences are 10% and 100%; in this situation, the relative difference seems more important.

At the same time, few would say that the difference between $1,000 and $3,000 is greater than the difference between $10,000 and $15,000. In this case, the absolute differences are $2,000 and $5,000, while the relative differences are 200% and 50%. Here we would probably judge the absolute difference to be the more important.

There is no mathematical preference for one over the other, leaving us free to choose which one to focus on. The first to suggest the use of absolute difference was James Dean, professor of Mathematics and Astronomy at Dartmouth College. In 1832, Dean sent a letter to Massachusetts senator Daniel Webster, a former student, arguing that future apportionments should try to reduce the absolute difference in the district sizes. Unfortunately, Webster ignored the suggestion of his former teacher and proposed (and ultimately passed) a method using a modified divisor that ignored any measure of malapportionment.

Relative difference suffered a similar fate, at least at first. During the 1910 apportionment debates, Joseph A. Hill, chief statistician of the Division of Research and Results for the Bureau of the Census, suggested that relative differences be the key measure of malapportionment. Hill's proposal also failed. But Congress's request that the National Academy of Sciences consider the apportionment problem gave relative difference a second chance.

To bolster support for the use of relative difference, Hill consulted Edward V. Huntington, a mathematician at Harvard University (but not a member of the panel). Huntington pointed out an important advantage to Hill's approach. Remember that A, B are the populations of two states, with a, b representatives.

Suppose we find an apportionment that minimized the absolute difference between district sizes A/a and B/b. Huntington noted that this would not, in general, minimize the absolute difference between congressmen per person a/A and b/B, nor the absolute difference between the ratio of congressmen and the ratio of populations a/b and A/B. However, an apportionment that minimized the relative difference between district sizes A/a and B/b would simultaneously minimize the relative difference between the congressmen per person and the relative difference between the ratio of congressmen and the ratio of populations. In this sense, a focus on relative difference insulates us from changing views on how malapportionment should be measured.

Of course, whether we choose to minimize the absolute or relative difference, we must still solve the problem of finding the apportionment. This seems to be a tremendously difficult problem: suppose we have an apportionment of congressional seats among the 50 states. To determine whether we should transfer seats between states, we must evaluate every possible transfer of seats between two states. If we do find a transfer that reduces the malapportionment, we must again examine every possible transfer among the new apportionments. How can this be done in a timely manner?

As it turns out, there is a simple and easy solution, whether we want to reduce absolute difference (Dean's proposal) or relative difference (Hill's proposal). Again, suppose we have two states with populations A, B, and that the states are currently allocated a, b representatives. Suppose we also have one congressional seat to allocate. If we want to reduce the absolute difference in district size, we should calculate $\dfrac{A}{a(a+1)}$ and $\dfrac{B}{b(b+1)}$, and whichever is larger corresponds to the state that should receive the extra congressional seat. Or, if we want to reduce the relative difference in district size, whichever of $\dfrac{A}{\sqrt{a(a+1)}}$ and $\dfrac{B}{\sqrt{b(b+1)}}$ is greater corresponds to the state that should receive the additional representative.

Consider for example the final apportionment following the 1790 census, and suppose we wanted to add one congressman to either Delaware (population 55,540 and 1 representative) or Virginia (population 630,560 and 19 representatives). If we wanted to minimize the absolute difference, we would compute $\dfrac{55{,}540}{1(1+1)} = 27{,}770$ and $\dfrac{630{,}560}{19(19+1)} \approx 1659$. Since the first is larger than the second, this means we should give the extra seat to Delaware. Equivalently, if we gave the extra seat to Virginia, we would reduce the absolute disparity by transferring the seat from Virginia to Delaware. Since Delaware would get the seat in any case, we may as well assign it to Delaware in the first place.

Likewise, if we wanted to minimize the relative disparity, we would compute $\dfrac{55,540}{\sqrt{1(1+1)}} \approx 39,273$ and $\dfrac{630,560}{\sqrt{19(19+1)}} \approx 32,347$. Again, since the first difference is larger than the second, we should give the extra seat to Delaware. And if we erred and gave the seat to Virginia, we would reduce the comparative disparity by transferring the seat to Delaware.

We can allocate all 435 seats as follows. Suppose we want to minimize the relative difference (Hill's proposal). For every state, compute the value $\dfrac{A}{\sqrt{a(a+1)}}$ for all values of a, from $a=1$ to $a=435$ (the maximum number of seats a state can have, though in practice we don't need to go so far). For example, Delaware would have, for the 1790 census, $A=55,540$. We would then compute $\dfrac{55,540}{\sqrt{1(1+1)}} \approx 39,273$, $\dfrac{55,540}{\sqrt{2(2+1)}} \approx 22,674$, $\dfrac{55,540}{\sqrt{3(3+1)}} \approx 16,033$, and so on. We would do the same for all other states, and then put the values in order, from highest to lowest, in what is referred to as a *priority list*. For the 1790 census data, the priority list would be:

Virginia: 445,873
Massachusetts: 336,107
Pennsylvania: 306,092
Virginia: 257,425
North Carolina: 249,979
New York: 234,469

and so on.

To perform the apportionment, we begin by giving each state its constitutionally mandated 1 representative, which allocates 15 representatives. Then we simply go down the list, assigning the seats to the states in the order they appear on the priority list. So the sixteenth representative would go to Virginia; the seventeenth to Massachusetts; the eighteenth to Pennsylvania; the nineteenth to Virginia; the twentieth to North Carolina, and so on. Because it is based on Hill's argument for relative difference and Huntington's algorithm, this method of apportionment is called the *Huntington-Hill method*.

Note that this method is independent of the actual number of seats to be assigned: if we want to assign more seats, we keep going down the list. Therefore the Alabama Paradox cannot occur, since if we want to assign additional representatives, we go further down the list. Nor can the population paradox occur, since a decrease in a state's population results in a decrease of all its priority values.

What if we wanted to minimize the absolute difference (the Dean method)? We would then compute $\dfrac{A}{a(a+1)}$ for all states; again, we would put the values in order in a priority list, which would be:

Virginia: 315,280
Massachusetts: 237,664
Pennsylvania: 216,440
North Carolina: 176,762
New York: 165,795

Notice that under the Huntington-Hill method, Virginia actually receives its third seat *before* New York and North Carolina receive a second seat; but under the Dean method, New York, North Carolina, Maryland, and Connecticut *all* receive their second seats before Virginia gets a third seat. Since we eventually run out seats, it is possible for a state to receive a seat under the Huntington-Hill method that it would not receive under the Dean method, and vice versa. In fact, if we apportioned 105 representatives using the Huntington-Hill method, Virginia would receive 18 seats, but under the Dean method, it would receive only 12! Clearly it is essential to decide which is more important: the absolute difference or the relative difference.

Congress temporized. In the wake of the failed 1920 apportionment, Connecticut Republican E. Hart Fenn sponsored a bill that permanently limited the size of the House to 435 members, apportioned according to the Webster method. The bill eventually passed. However, an amendment required reporting the apportionments produced by two other methods: Huntington-Hill's, and the method of major fractions (a variation on Webster's method). President Hoover signed the bill into law on June 19, 1930.

Both the Huntington-Hill and Webster methods produced identical apportionments for the 1930 census. But in 1940, they differed: the Webster method gave Arkansas 6 representatives and Michigan 18, while the Huntington-Hill method gave Arkansas 1 additional seat at the expense of Michigan. Mathematically, the Huntington-Hill method produces an apportionment that minimizes the relative difference between district size and thus results in a "fair-as-possible" apportionment, while the Webster method makes no claim to reduce any malapportionment. If our goal is to reduce malapportionment, the Huntington-Hill method, insofar as it minimizes the relative difference in district size, should be regarded as superior to the Webster method.

Of course, what is mathematically sensible must also be politically palatable. As it turned out, Democrats controlled both houses of Congress and the presidency in 1940, and Arkansas leaned Democratic while Michigan leaned Republican. For once, the rational solution was politically feasible, and President Franklin Roosevelt signed Public Law 291, mandating the use of the Huntington-Hill method, on November 15, 1941. The Huntington-Hill method has been used in all apportionments since.

Embracing the Error

One problem with the current method of apportionment is that the slightest error in determining a state's population could change its position on the priority list, and cost it a congressional seat. For example, suppose we have just two states A and B with recorded populations of 1,000,000 and 3,162,278 respectively. If we apportioned six congressmen among them using the Huntington-Hill method, state B would receive five congressmen and state A would receive one.

But suppose one census form from state A was misplaced, and that one person in state A was not counted. State A's actual population would be 1,000,001. Using this instead of 1,000,000 as the population of state A would change the apportionment: B would receive four congressmen and A would receive two. We would observe a similar alteration in congressional apportionment if a misread census form caused state B's population to be one greater than it actually was (in other words, for the true population of state B to be 3,162,277). The Huntington-Hill method is more sensitive to such errors than the Dean method: in particular, it takes a smaller error in a population count to alter an apportionment made under the Huntington-Hill method than under the Dean method.[4]

---

4. In 2000, under the Huntington-Hill method, Utah needed 856 more residents out of 2,236,714 to win the last congressional seat. Had the Dean method been used, New Mexico would have won the last seat with 1,143 more residents out of 1,823,821: both a larger absolute and relative error. If "preventing lawsuits" is a constitutional imperative, this could be used to argue for the superiority of the Dean method!

We could spend even more time, money, and effort to make the census even more accurate. But suppose we used sampling and the statistical margin of error. Remember that the margin of error does not reflect actual mistakes in the census process, but rather the probability that the sample mean and standard deviation differ from the population mean and standard deviation.

Suppose we take a large enough sample to obtain state populations to within 0.5%. Then state A's population would be given as $1,000,000 \pm 5,000$, and state B's population as $3,162,278 \pm 15,811$. We could then use the margin of error to produce a priority *interval*, and assign representatives to states with the highest priority intervals. For example, state B's first priority interval would be $\dfrac{3,162,278 \pm 15,811}{\sqrt{1(1+1)}}$, or 2,224,888.152 to 2,247,248.283, while state A's first priority interval would be $\dfrac{1,000,000 \pm 5,000}{\sqrt{1(1+1)}}$, ranging from 703,571.247 to 710,642.315. Since every value in B's priority interval is higher than every value in A's priority interval, there is no question that B should receive the third representative assigned. Likewise, B would receive the fourth and fifth.

What about the sixth and last representative? In this case state B, with four seats already, would have a priority interval of $\dfrac{3,162,278 \pm 15,811}{\sqrt{4(4+1)}}$, or 703,571.410 to 710,642.304. Notice that the priority interval for B's fifth seat overlaps the priority interval for A's second seat. Statistically speaking, this means that there is a tie for the last seat: there is insufficient evidence to decide which state should receive it. The sensible thing to do is to give *both* states an additional seat. Thus B receives one more seat, for a total of 5; and A *also* receives one more seat, for a total of 2.

Note that under this system, we could have assigned five seats (four to B and one to A) or seven seats (five to B and two to A), but since the sixth and seventh seats were assigned simultaneously, it would have been impossible to assign exactly six seats.

This runs afoul of the Apportionment Act of 1911, which set the size of the House of Representatives at exactly 435 congressmen. Although we may not want more Congressmen, limiting their numbers means that there will always be the potential for a *Utah v. Evans* situation, where the slightest error in the census count or disagreement over who should be counted could alter the congressional apportionments for two or more states. The simplest alteration of the Apportionment Act would be to use priority intervals and assign at least 435 seats. If two (or more) states' priority intervals overlap, then seats would be simultaneously assigned to these states. Under this system, a state would receive at least as many seats as it would under the current system. The only important difference will occur when assigning the last seat.

The most important advantage to this system is that any actual errors in the conduct of the census must have a magnitude greater than the statistical margin of error to be politically relevant. Under the current system, as trivial an error as miscounting a single person could alter a state's congressional delegation. In contrast, a state whose priority *interval* misses the cutoff for the last group of seats would have to argue for the existence of an error greater than the listed margin of error. The first type of error could easily go unnoticed; the second would require some catastrophic event.

Such events have occurred. For example, in 1850 the records for several California counties were lost at sea. The remaining records gave California a population of 93,000. However, Superintendent of the Census Joseph Kennedy imputed the missing data and reported an estimated population of 165,000. The Hamilton method had just been reinstated as the means for calculating the number of representatives due a state, and based on its population of 165,000, California was slightly higher on the priority list than South Carolina; thus it would receive one of the additional seats apportioned. However, Kennedy noted that if the California population was 135,000, the additional seat would go to South Carolina instead. Congress chose to use the 165,000 figure, presumably not believing so large an error was possible.

Using priority intervals also means that there is no need to conduct a door-to-door census, since the margin of error is incorporated into the apportionment process. The only question is how large we want the margin of error to be. A useful guideline might be the following: depending on the margin of error and the number of seats already allocated, it is possible for states' priority intervals to appear multiple times in a tie situation; a state's priority interval could even overlap itself. We can avoid this situation by choosing a sufficiently small margin of error.

For example, the original plan for Census 2000 would have resulted in a sample consisting of at least 90% of the U.S. population, which, under the most wildly exaggerated assumptions about the magnitude of the error, would still allow the U.S. population to be determined to within about 0.15%.

At this margin of error, a state's priority interval would not overlap itself until it had already been assigned 383 congressional seats. This is about six times as many seats as the most populous state (California) is currently assigned; in fact, were any state to actually receive 383 additional seats, there would only be 2 seats left to apportion among the remaining 49 states! It follows that this margin of error more than suffices to use priority intervals for the apportionment problem.

Weighting for a Fair Vote

> We have barely crossed the threshold in exploring the variety of devices
> which may be employed, consistent with the constitutional mandate of
> one man-one vote, to assure that the points of view of the smaller towns
> in the county will be heard on the Board of Supervisors.
> —STANLEY FULD, CHIEF JUDGE, NEW YORK COURT OF APPEALS,
> *IANNUCCI V. BOARD OF SUPERVISORS* (1967)

Following the 2010 census, the 309,183,463 persons eligible for representation
in the United States would be apportioned 435 congressmen. On average,
each congressman would have 710,767 constituents. But as *Utah v. Evans* and
similar cases show, states may gain or lose seats with very small changes in
population. Montana's 2010 population of 994,416 was not high enough to
qualify it for a second congressional seat, so its single congressman represents the entire population of the state. Meanwhile, Rhode Island's 2010 population of 1,055,247 was just high enough to win it a second seat, so each of its
two congressmen represents 527,624 persons.

The Apportionment Act of 1911 limited Congress to 435 members, but the
U.S. population in 1911 was one-third its present size. As a result, a U.S. congressman now represents more people than their counterparts in almost
every other democracy. For example, a member of the British House of Commons represents about 100,000 constituents, while members of the German
Bundestag represent about 200,000 persons. As early as the presidencies of
George Washington and Thomas Jefferson, the size of congressional districts
was in question: representatives could not truly understand all the concerns
of constituents in a too-large district.

The obvious solution is to increase the size of the House. In 2010, a group
of voters from Mississippi, Delaware, South Dakota, Montana, and Utah (the
states with the most populous congressional districts in the United States)

sued to overturn the limit set in 1911. But in *Clemons v. Department of Commerce* (2010), the District Court for the Northern District of Mississippi ruled that the size of Congress was a political issue, which should be addressed through legislative, not judicial, action.

Suppose Congress could be persuaded to increase the number of representatives. How large should the House of Representatives be? The only specific reference to congressional district size in the Constitution is the proviso that there should be no more than one representative for every 30,000 persons. In *Wendelken v. Bureau of the Census* (1983), Martin Wendelken of New Jersey argued that this *required* one representative per 30,000 persons. Such an allocation would produce (in 1983) a congress with about 7,000 members. Wendelken's suit, like Clemons's, was dismissed by the District Court.

A Congress with 7,000 members would likely be too unwieldy to be useful. The British House of Commons, German *Bundestag*, and India's *Lokh Sabah* each have around 600 members, so we might take 600 as a reasonable size for a national legislature. However, simply increasing the size of the House of Representatives is not enough: the law should take into account future population growth.

One possible approach is the *Wyoming Rule*. This is essentially a divisor rule that sets the ideal size of a congressional district to that of the least populous state (which, since 1990, has been Wyoming). Thus the 2010 census gave Wyoming an apportionment population of 568,300. If the national population of 309,183,463 were divided into 568,300-person congressional districts, we would need 544 representatives.

Unfortunately, even an increase to 544 members would do almost nothing to reduce the disparities in district sizes: If we used the Huntington-Hill method of apportionment, Alaska would have the most populous congressional district, with 721,523 persons, while South Dakota would have the two smallest, with 409,881. This produces an absolute difference of 311,642 persons and a relative difference of 76%!

Weighted Voting

Another possibility is to use a *weighted voting system*, by which every representative casts more than one vote. There are about 15,000 readily available examples of weighted voting systems, since this is how every stockholder-owned company is run. As in the federal government, day-to-day operations are handled by non-elected employees generally chosen on the basis of merit; however, major strategic decisions are made by elected officials. The main difference is that the electors are the shareholders, who cast a number of votes equal to the number of shares they own.

In 1917, Nassau County in New York adopted a weighted voting system. The county consists of three towns (Hempstead, North Hempstead, and Oyster Bay) and two cities (Long Beach and Glen Cove), and the city or town supervisors had a number of votes computed by dividing the number of constituents by 10,000 and rounding down. Hempstead actually had two supervisors, so the county charter mandated that the two supervisors split the votes between them. Based on the 1960 census figure, this gave the two Hempstead supervisors 36 votes apiece; Oyster Bay had 28; North Hempstead, 21; Glen Cove and Long Beach, 2 votes apiece. As there are 125 votes altogether, a majority requires 63 votes.

There is a problem with this allocation: the town of Hempstead by itself controls a majority of votes; hence the two supervisors, working together, could pass any measure. The county charter forbids this; when a similar situation occurred in 1937, the county attorney simply reduced the number of votes given to these supervisors, but kept the number required to pass a measure the same. The county clerk, citing the 1937 precedent, reduced the representation of Hempstead, producing the final allocation shown in table 4. With a new total of 115, a majority required 58 votes. However, the people of North Hempstead, Glen Cove, and Long Beach noticed a peculiar pattern: their votes never seemed to matter.

Table 4.   1965 Weighting for Nassau County Board of Supervisors

District	Votes
Hempstead 1	31
Hempstead 2	31
Oyster Bay	28
North Hempstead	21
Glen Cove	2
Long Beach	2

New York lawyer John F. Banzhaf III considered the problem as follows. What matters is not that a person *has* a vote, but rather that each person's vote could make a difference. Consider a voting system with three electors *A*, *B*, and *C*, who cast 3, 1, and 1 vote respectively, where a total of 4 votes are needed to pass a measure. A *winning coalition* is a group of one or more electors who have enough votes to pass a measure. For example, {*A*, *B*} is a winning coalition, since *A* casts 3 votes and *B* casts 1. Another winning coalition is {*A*, *B*, *C*}.

Suppose one member of a winning coalition *defects* and leaves the coalition. This may or may not change a winning coalition into a losing coalition. For example, in the winning coalition {*A*, *B*}, the departure of *B* will turn this into a losing coalition consisting of *A* alone. On the other hand, if *B* leaves the winning coalition {*A*, *B*, *C*}, the remaining members of the coalition still have enough votes to pass a measure. If an elector's departure from a winning coalition turns it into a losing coalition, we say that the elector is *pivotal* or *critical*.

Banzhaf's insight was that we could measure the importance of a voter by how many times he or she was pivotal. Thus elector *A* will be pivotal in three winning coalitions: {*A*, *B*}, {*A*, *C*}, and {*A*, *B*, *C*}. On the other hand, elector *B* will only be pivotal in one winning coalition, {*A*, *B*}; likewise for elector *C*. Since there are a total of five pivotal electors (A three times, *B* and *C* once apiece), the *normalized Banzhaf index* for *A* will be 3/5, for *B* and *C* it will be 1/5.[1]

Computing the Banzhaf index is, in most cases, a tedious task best left to a computer. However, in some cases we needn't calculate the Banzhaf index for every voter; we can focus on those we are interested in. Of particular interest are the *dummy voters*—those who are *never* pivotal. The significance of the dummy voter is the following. Suppose a voter is part of a winning coalition. Since a dummy voter is never pivotal, the coalition will remain winning even if he or she leaves the coalition. By a similar argument, a dummy voter cannot join a losing coalition and turn it into a winning coalition. Thus, winning

---

1. Like most good ideas, several people came to the same conclusion independently. Mathematician Lionel Penrose first suggested the measure in 1946, and sociologist James Coleman in 1971. The fact that a mathematician, lawyer, and sociologist independently arrived at similar conclusions about how to measure voting power argues for a greater interaction among the disciplines of mathematics, law, and sociology.

coalitions need not address the interests of dummy voters, and losing coalitions have nothing to gain by recruiting them.[2]

Consider the North Hempstead supervisor, who casts 21 votes. Since 58 votes are needed to pass a measure, this elector will be pivotal if he or she is part of a coalition that casts between 58 and 78 votes: 58 since the coalition must be winning, and 78 since the coalition will be losing without these 21 votes.

The crucial question is this: How many such coalitions are there? The answer is *none*. This means there is no winning coalition in which North Hempstead's supervisor is pivotal, and so the power of this supervisor, under the Banzhaf index, is 0. The same holds for the supervisors from Glen Cove and Long Beach. We can compute the Banzhaf index for all of the supervisors: those of Hempstead and Oyster Bay all have index 1/3, while those of North Hempstead, Long Beach, and Glen Cove have index 0.

There are other measures of voting power; the most common alternative to the Banzhaf index is the *Shapley-Shubik index*, invented by Lloyd Shapley and Martin Shubik in 1954. The main difference between the Banzhaf index and the Shapley-Shubik index is that the latter focuses on growing a coalition by adding electors; the elector whose vote turns the coalition into a winning coalition is the pivotal voter.

Again, consider our three-elector system with electors *A*, *B*, and *C* who respectively cast 3, 1, and 1 votes apiece, with 4 votes needed to pass a measure. Imagine that we take a vote by having each elector vote in some sequence. There are six ways the electors can vote:

ABC   ACB   BAC   BCA   CAB   CBA

Imagine any given elector, say *B*. There are two possibilities to consider. First, *B* might be the sponsor of a measure. It should be clear that if voting has any meaning at all, *B* by itself should never be pivotal. Thus *B* needs to find allies.

As there are two other electors, *B* can approach either one and in principle win their support. If *B* approaches *A* first, then winning their support gives *B* enough votes to pass a measure. Thus *A* is pivotal in building a coalition to support *B*'s proposal. On the other hand, if *B* approaches *C* first, then even if

---

2. "Dumb" originally meant (and still does, among specialists) "unable to speak." A dummy voter is not an unintelligent one, but one who (metaphorically, at least) has no voice.

B wins *C*'s support, they do not have enough votes to pass a measure: the two of them will still need to gain *A*'s support.

The first situation corresponds to the voting order *BAC*, while the second corresponds to the voting order *BCA*. In both cases, *A* is pivotal, turning a losing coalition into a winning coalition. The Shapley-Shubik index measures how often this occurs.

In our example, we see that *A* will be the pivotal voter in four coalitions: *BAC*, *BCA*, *CAB*, *CBA*. Meanwhile *B* and *C* will be the pivotal voters in one coalition each: *ABC* and *ACB*. *A* is the pivotal voter in $4/6 = 2/3$ of the coalitions, while *B* and *C* are the pivotal voters in 1/6 of the coalitions, and their Shapley-Shubik values will be 2/3, 1/6, and 1/6 respectively.

The two measures give different values for voting power, but this shouldn't be too surprising, as they measure different things. Roughly speaking, the Banzhaf index measures the importance of a member of a coalition and, to some extent, how much effort must be made to retain them. Meanwhile, the Shapely-Shubik index measures the importance of a non-member and, to some extent, how much effort should be made to recruit them.

While the problem with the Nassau County apportionment was evident from Banzhaf's analysis, it was not clear what should be done about it. In fact, several citizens from Hempstead launched their own lawsuit, holding that the *reduction* of the votes cast by their supervisors was a violation of *their* constitutional right to representation.

As the case wound its way through the legal system, the Court of Appeals of the state of New York took a bold step and established Banzhaf's approach as the method by which weighted voting systems should be assessed.[3] In *Ianucci v. Board of Supervisors* (1967), which concerned the weighted voting systems in Saratoga and Washington County, Chief Judge Stanley Fuld wrote: "The principle of one man-one vote is violated, however, when the power of a representative to affect the passage of legislation by his vote, rather than by influencing his colleagues, does not roughly correspond to the proportion of the population in his constituency" (20 N.Y.2d 244, p. 252). Judge Fuld gave an example where a legislator who represented 60% of the voters and who

---

3. The Court of Appeals is the highest of the state courts of New York. Confusingly, the Supreme Courts of New York are the highest courts in the counties, and are inferior to the Court of Appeals.

cast 60% of the vote would, in fact, have 100% of the voting power whenever a simple majority is required.

Fuld went on to suggest that each legislator in a weighted voting plan should be able to cast the decisive vote in proportion to the number of their constituents. Thus a legislator who represents 60% of the population should be the pivotal voter 60% of the time. This leads to the *inverse Banzhaf problem*: find the appropriate voting weights so that each elector is decisive in a specified fraction of the votes.

On the one hand, this suggests that the county clerk's decision to reduce the number of votes cast by the Hempstead supervisors was warranted, since otherwise the Hempstead electors would be pivotal 100% of the time with 57% of the population, while the others, with 43% of the population between them, would never be pivotal. On the other hand, no *single* legislator cast more than a majority of the votes, so Justice Fuld's ruling in *Ianucci* could be held irrelevant. The Nassau County Supreme Court took the latter viewpoint and struck down the apportionment in *Franklin v. Mandeville* (1968). The court directed the Board of Supervisors to come up with a new weighting scheme within six months, though it later allowed them to defer implementation until after the 1970 census results were certified.

To solve the inverse Banzhaf problem, the Board of Supervisors used a computer analysis of more than 2,000 plans, and eventually settled on the following. First, the six members would have weights as follows: Hempstead, 35 votes apiece; Oyster Bay, 32; North Hempstead, 23; Long Beach, 3; Glen Cove, 2. Next, in order to pass a measure, a supermajority of 71 votes, not 66, was required (and measures requiring a "two-thirds" vote required a total of 92, not 87, votes). Under this new plan with the higher quota, the two supervisors for Hempstead represented 56% of the population and cast 55% of the decisive votes. Again, the failure of the weights to be in proportion to the populations led to a lawsuit, but in *Franklin v. Krause* (1973), the Court of Appeals upheld it, effectively establishing the Banzhaf index as the measure for weighted voting fairness.

"The Haze of Slogans and Numerology"

Unfortunately, weighted voting is so uncommon outside of New York State that lawmakers and judges rarely consider its effects. When Indiana redrew its state legislative districts following the 1970 census, it established several multimember districts that elected two or more at-large representatives. For example, voters in Marion County sent 15 representatives to the state legislature.

If the representatives vote as a bloc, they are effectively a single "legislator" with 15 votes. However, since the *number* of votes this "legislator" casts is proportional to the population they represent, it follows that there can be a substantial disparity between the population represented and the power in the legislature. *Chavis v. Whitcomb* (1971) argued that the existence of a multi-member district disadvantaged voters in single-member districts. The Supreme Court was not convinced, and let the districts stand as they were drawn.

Though Justice John Marshall Harlan supported the decision, he wrote a rather caustic opinion expressing a view that his fellow justices lacked "a coherent and realistic notion of what is meant by 'voting power'" (403 U.S. 124, p. 168). Harlan, who was even then dying of spinal cancer (and perhaps felt no need to suffer fools gladly), pointed to the Banzhaf index as an example of how a theoretical model can shed light on decisions about apportionment and, in particular, provide a way to measure the voting power across districts. He admitted that the results depend sensitively on the assumptions made by the model, so that wholehearted acceptance would be unreasonable; nevertheless, "[the Court] neither suggests an alternative nor considers the consequences of its inability to measure what it purports to be equalizing. Instead it becomes enmeshed in the haze of slogans and numerology which for 10 years has obscured its vision in this field" (403 U.S. 124, p. 169).

Eighteen years later, the Court remained enmeshed in the haze of slogans and numerology. But rather than rely on one of the quantifiable methods of measuring voting power, the Court decided to reject such an approach entirely.

In the 1980s, the New York City Board of Estimates made decisions about property management, city contracts, zoning, and budgets. The Board consisted of a number of officials elected by citywide vote, as well as the presidents of the five boroughs: Manhattan, Staten Island, Queens, the Bronx, and Brooklyn.

Based on the 1980 census, Brooklyn had a population of 2,230,936, while Staten Island had a population of 352,121. Since both had one representative on the Board of Estimates, this seemed to strike at the heart of equal representation. The Supreme Court heard *Board of Estimates v. Morris* (1989) beginning on December 7, 1988. The Supreme Court eventually decided that the

Board of Estimates was malapportioned, and ruled unconstitutional the section of the city charter that established it. Subsequently, the duties of the Board of Estimates were taken over by the City Council, and the board no longer exists.

The Court based its decision on the gross population disparities among the boroughs, and *not* on any notion of voting power. In fact, while the Court heard an argument based on the Banzhaf index, it derided the analysis as too theoretical to be relevant.

There are certain assumptions that we must make in order to compute the Banzhaf index. To return to the original Nassau County example, the two supervisors for Hempstead form a winning coalition, and either one is pivotal. However, if political alliances meant that these the two supervisors *never* formed a coalition, then we cannot count them as pivotal: the required electoral situation never exists.

The problem is that the Banzhaf index measures the probability that an elector is the decisive voter, under the assumption that all voters are as likely to support a measure as to oppose it. But, as Justice Harlan pointed out in *Chavis*, even a slight predilection in one direction will substantially alter this probability. In this sense, the Banzhaf index does not measure actual voting power, but rather theoretical voting power.

Yet it's not clear that this objection is reasonable. For example, the 26th Amendment guarantees every U.S. citizen over the age of 18 the right to vote. This theoretical right to vote is all that matters; the fact that some groups vote more reliably than others is constitutionally irrelevant. By a similar argument, the actual pattern of voting should be irrelevant to the question of the relative power of different voters. Justice Fuld made an oblique reference to this in *Ianucci*: it is the *theoretical* possibility of casting the pivotal vote that matters.

Moreover, as Justice Harlan also noted, purely theoretical voting indices can still shed light on the question of voting power. This suggests a different strategy. Rather than presenting the Banzhaf indices as evidence that the weighted voting system is unfair, it would be better to find an obviously unfair weighted voting system with the same Banzhaf values.

As an example, consider the Nassau County case, but suppose that instead of assigning the votes as in table 4, we give the two Hempstead supervisors and the Oyster Bay supervisor one vote, and the other three supervisors zero votes. It should be obvious that this system is unfair.

What's not obvious is the following. Consider any coalition in the original system, say the second Hempstead, Oyster Bay, and Long Beach supervisors, who together would have $31 + 28 + 2 = 61$ votes. This would be a winning coalition in the new system, with $1 + 1 + 0 = 2$ votes (out of 3 total votes). Or consider the coalition of North Hempstead, Glen Cove, and Long Beach, which has $21 + 2 + 2 = 25$ votes in the old system, and is thus a losing coalition; in the new system, this coalition will have 0 votes, and will still be a losing coalition. As it turns out, if a coalition is winning in the original system, it will be winning in the new system, and vice versa; and if a coalition is losing in the original system, it will still be losing in the new system, and vice versa. We say the two systems are *mathematically equivalent*. From a political point of view, the outcomes of any vote in the original system will be the same in the new system—but in the new system, it is far more obvious that three of the electors are irrelevant.

Weighted Voting in Congress

The persistent lawsuits over the weighting of supervisor votes in Nassau County eventually led them to abandon weighted voting entirely, though it remains in use in a number of New York counties, including Saratoga and Washington County (the focus of the *Ianucci* suit). Moreover, the Supreme Court has not yet ruled against it, so there is no a priori reason to believe it unconstitutional.

In fact, one interpretation of the Constitution holds that omissions are just as significant as inclusions. In particular, the Constitution specifies that each senator casts one vote, but does not specify how many votes should be cast by each representative. The omission suggests that it would be constitutionally permissible to allow weighted voting in the House of Representatives.

We might consider Fuld's suggestion—that the power of a representative be proportional to the number of persons each represents. In Nassau County, trying to engineer this required a computerized analysis of thousands of voting schemes—and in this case, only six electors were involved. Trying to engineer a similar solution for Congress, with 435 representatives, would be much more difficult.

However, we might not need to go this far. The population differences among congressional districts are not nearly as large as the population differences among the cities of Nassau County, so the voting weights should be roughly equivalent. Under these circumstances, the Banzhaf index is roughly proportional to the square root of the number of votes cast by an elector. It follows that if each congressman casts a number of votes equal to the square root of the population of their district, their voting power would be roughly in proportion to the number of persons they represent.

This is the basis of the *Jagiellonian compromise*, a proposal by physicist Karol Zyczkowski and mathematician Wojciech Slomczynski, both from the Jagiellonian University in Krakow, to weight the votes of the member states in the European Union according to the square roots of their respective populations. If we applied the Jagiellonian Compromise to Congress, then Montana's single congressman, whose district of 994,416 people (based on the 2010 census) is the largest in the country, would cast 997 votes. Meanwhile each of Rhode Island's two congressmen, whose districts of 527,624 are the smallest in the country, would cast 726 votes.

There is, of course, the risk that one or more congressmen might be dummy voters. However, this occurred in Nassau County largely because of the great disparity between the supervisors with the most votes (the two Hempstead supervisors, at 31), and a large gap between the number of votes cast by the two lightest voters, Long Beach and Glen Cove (at 2 apiece) and the next heaviest elector, North Hempstead (at 21).

In contrast, there would be a much smaller range between the number of votes cast by each member of Congress: from Montana's congressman with 997 votes, to Rhode Island's two congressmen with 726 votes. Moreover, the intermediate numbers are well represented: thus Wyoming's representative would cast 754 votes; Nebraska's representatives would cast 781; West Virginia's would cast 787; and so on. This would be a simple way to implement weighted voting in Congress and ensure that each voter, regardless of the size of their congressional district, has a roughly equal weight in Congress.

The Impossibility of Democracy

It is generally believed . . . that in an election by voting, the plurality of
votes always indicates the views of the electors . . . But I will make clear
that this opinion, which is true in the cases where the election is
between two candidates only, can lead to an error in all other cases.

—JEAN CHARLES DE BORDA,
MEMOIR ON ELECTIONS BY VOTING (1770)

In November 2012, voters in California's 40th congressional district faced an
unusual choice: between incumbent Lucille Roybal-Allard and challenger
David Sanchez. What made this unusual is that both Roybal-Allard and
Sanchez were Democrats. No Republican candidates appeared on the ballots
for the 40th district. An analogous situation held in the 31st district, where
two Republican candidates, Bob Dutton and Gary Miller, faced each other
and no Democrats appeared on the ballot. Several other districts across the
state saw similar competitions between two candidates belonging to the same
political party.

The situation in California is an unusual solution to a common problem:
how should we elect public officials? In most of the United States, elections
are decided by *plurality*: voters are presented with a list of candidates; they
vote for one candidate; and the candidate with the most votes wins.

Plurality voting has been criticized on a number of grounds. One of the
more common objections is that two candidates may split like-minded voters
between them, allowing for the victory of a third candidate. For example,
in 1912, Theodore Roosevelt broke from the Republican Party and ran as an
independent. Voters had a choice between an unpopular Republican president (Taft); a popular ex-Republican ex-president (Roosevelt); and a Democratic challenger (Wilson). It is commonly believed that Roosevelt's candidacy split the Republican vote, allowing Wilson to win the election.

More recently, the victory of George W. Bush in 2000 has been attributed to progressive voters splitting their support between Ralph Nader and Al Gore. And in the 2012 election cycle, Republicans in Maine, New York, Arizona, and other states were accused of supporting Green Party candidates *because* they would draw votes away from their Democratic competitors.

The Constitution only specifies that elections must be held, but it leaves open the nature of the voting process. We can abandon plurality and replace it with something different—but can we replace it with something better?

Condorcet Winners and Losers

In the latter half of the eighteenth century, France suffered from policies that exempted the upper classes from taxation; stock market crashes that wiped out the middle class; bad weather that produced several years of poor harvests; and involvement in several pointless and costly foreign wars. As a result, the French government was bankrupt. Money might be raised if members of the nobility allowed themselves to be taxed, but they successfully blocked every attempt to exact reform through legislative action.

Mathematical politics was born against this backdrop, and, like so many things, originated because of a disagreement between two persons over how a problem should be solved. And like so many things, the problem at issue was not one of the great questions of the day, but instead involved something that seemed rather trivial: how should members of the French Academy of Sciences be elected?

In 1770, Jean Charles Borda (1733–1799), a cavalry officer whose study of ballistics earned him membership in the academy, proposed a method now known as a *Borda count.* Rather than having each voter choose one candidate, Borda suggested that the voters rank all the candidates, from best to worst. Then for every last-place vote a candidate won, he or she received one point; for every next-to-last-place vote, two points; for every third-to-last place, three points; and so on. The candidate who received the largest point total won the election.

In modern terms, each voter submits a *preference list.* For example, if there are four candidates, x, y, z, w, a voter who prefers x to y, y to z, and z to w would have preference list $x > y > z > w$; in this case, we might say that this voter ranked x highest, y second-highest, z third-highest, and w last. We

Table 5. Example of an Election Profile

Ranking	Number of Ballots
$x > w > y > z$	150
$y > w > z > x$	140
$z > y > w > x$	130
$w > z > x > y$	100
All other rankings	0
Total	520

assume that the voter preferences satisfy *transitivity*, so that if $x > y$ and $y > z$, we can conclude that $x > z$.[1] It's convenient, though not absolutely necessary, if we disallow ties in the preference lists. The collection of the preference lists of all voters is known as the *election profile*.

Table 5 provides an example of an election profile in which 520 voters submitted ballots. The table shows that 150 voters submitted a ballot $x > w > y > z$, indicating they ranked x the highest, w second-highest, y third-highest, and z last; 140 voters submitted a ballot of $y > w > z > x$, indicating they ranked y highest, w second-highest, z third-highest, and x last; and so on. Note that other preference lists are possible, for example $x > y > z > w$, but in our example we'll assume that only the preference lists shown were submitted.

A *social welfare function* takes the election profile and returns a *societal ranking* of all the candidates. The ranking must satisfy *transitivity*. It's convenient, though not absolutely necessary, to disallow ties; in this case, the social welfare function is said to be *resolute*.[2]

For example, consider the Borda count. From the profile in table 5, we see that z received 150 last-place votes (from the voters with preference list $x > w > y > z$); 140 third-place votes, 100 second-place votes, and 130 first-place votes.

---

1. Nontransitive ranking is actually common. Consider a dessert menu: we may prefer cannoli over tiramisu, and tiramisu over cheesecake, but it does not follow that we prefer cannoli over cheesecake.

2. Resoluteness really means that if a tie occurs, there is some nonrandom means of resolving it from the election profile. For example, the 2009 Town Council election in Cave Creek, Arizona ended in a tie, 660 to 660. A resolute system would break the tie using some deterministic method. For example, it might give victory to the candidate who registered their candidacy first. In fact, the Cave Creek election was decided by drawing cards. While this is a legitimate means of resolving ties, the system produced is not resolute, since outcome of drawing a card is random, not deterministic.

Thus z would get 1 point for each of the 150 last-place votes, 2 points for each of the 140 third-place votes; 3 points for each of the 100 second-place votes; and 4 points for each of the first-place votes, for a total of $1 \times 150 + 2 \times 140 + 3 \times 100 + 4 \times 130 = 1,250$ points.

By a similar computation, we find that x received 1,070 points, y received a total of 1,350 points, and w received 1,530 points. Putting these point totals in order, we obtain the societal ranking of $w > y > z > x$. Since w has the highest societal ranking, the Borda count would declare w the winner of the election.

The Borda count drew the scrutiny of fellow academician Marie Jean Antoine Nicolas de Caritat, Marquis de Condorcet (1743–1794). Born into nobility in Ribemont (in northern France), Condorcet studied mathematics and was elected to the French Royal Academy of Sciences in 1769. His work made him the founding figure of both mathematical politics and the viewpoint that mathematics would inevitably lead to a better society.

Condorcet identified a key problem with the Borda count. Many political scientists and philosophers subscribe to the *majoritarian principle*: If a majority of the voters prefer one choice, that choice should be the winner (or, if there can be more than one winner, that choice should be one of the winners). In elections with just two candidates, such as California's congressional elections, the winning candidate will necessarily have a majority and the system will automatically satisfy the majoritarian principle. But if there are three or more candidates, it's possible that the candidate with the most votes will fail to win a majority.

Condorcet proposed that the majoritarian principle could be extended to multi-candidate races as follows. Suppose a candidate could defeat any of the other candidates in a two-way race. That candidate would be called the *Condorcet winner*.[3] The majoritarian principle implies that the Condorcet winner should win the election. If this always occurs, then the voting rule satisfies the *Condorcet winning condition*.

---

3. We state here, once, what should be stated multiple times in multiple contexts: almost no one names something after themselves. Borda did not call his method the Borda count, nor did Condorcet use the term Condorcet winning condition. But to avoid awkward constructions like "Condorcet proposed what is now known as the Condorcet winning condition . . ." we will hereafter identify the originator of a concept and the modern name without the observation that it *is* the modern name.

There's one problem: If there are more than two candidates, how can we determine who would win a race between just two of them? Mathematical politics was born when Condorcet realized that the election profile could be used to identify the winner of *any* two-candidate race, and in fact the winner of the election, using any voting rule whatsoever.

For example, suppose the elections discussed previously were decided by plurality, where each voter supports one candidate, and the candidate with the most votes wins. Voters with preference list $x > y > z > w$ feel that candidate x is the best of the four. If we assume that these voters will vote for the candidate they rate highest (an assumption we will examine later), then they will vote for x. Likewise, the voters with preference list $y > x > z > x$ will vote for y; the voters with preference list $z > y > w > x$ will vote for z; and the voters with preference list $w > z > x > y$ will vote for w. Based on the profile in table 5, we can determine the final vote count to be x, 150 votes; y, 140 votes; z, 130 votes; and w, 100 votes. Plurality will return the societal ranking $x > y > z > w$, and x would be the winner of the plurality vote.

But what if the election was between x and w only? According to the profile, 150 voters prefer x to w, namely those with preference list $x > w > y > z$. On the other hand, $140 + 130 + 100 = 370$ voters prefer w to x. So in a race between x and w, w would be the winner.

If we examine table 5, we see that y would win in a two-way race against x (with 270 votes to 250); y would also win in a two-way race against z (290 to 230) and w (270 to 250), which means y is the Condorcet winner. But x won the plurality vote and w won the Borda count, so both voting rules can violate the Condorcet winning condition.

Later writers have incorporated the *Condorcet losing condition*, which requires that if a Condorcet loser exists, that candidate must *not* be the winner of the election. If we look at the profile carefully, we see that x will lose a two-way election against y, z, or w, so x is the Condorcet loser. But remember that x would be the winner by plurality vote; thus plurality voting violates the Condorcet losing condition. Interestingly, the Borda count never violates the Condorcet losing condition, though this was not proven until 1976.

The Condorcet winning condition suggests a perfect voting method: Obtain the election profile, and declare the Condorcet winner the victor. Unfortunately, not all election profiles have Condorcet winners. For example, suppose we have three candidates a, b, and c and 150 voters, with 50 voters having preference list $a > b > c$; 50 voters having preference list $b > c > a$; and 50 voters having preference list $c > a > b$. In a contest between a and b, a will win 100 to 50, but in a contest between a and c, c will win. The same is true for all candidates: each of them will win one of the possible two-candidate elections and lose the other.

Blanket Primaries

What voting rule shall we use if there is no Condorcet winner? One possibility is plurality, *if* we can remove its objectionable features. Thus far, there are three real objections: it violates the Condorcet winning condition, it violates the Condorcet losing condition, and also violates the majoritarian principle. These can be avoided if there are only two choices, as occurs in certain types of elections, such as referendums or recall votes. For this type of election, plurality is not only suitable but, for reasons discussed later, the ideal form of voting.

In contrast, there are generally more than two candidates for any elected office. If we wish to use plurality and avoid its more unpleasant features, we must find a way to limit the voter's choices to just two candidates. This could be done through state boards of election, which not only have the authority to limit which candidates appear on the ballot, but (following *Burdick v. Takushi* 1992) can even prohibit write-in candidates.

How can we limit a voter's choices to one of two candidates? The *primary system* suggests one way. In most states, every registered political party can put its nominee on the ballot in the general election. These nominees are chosen in primary elections, where voters choose from candidates seeking the nomination of a particular political party. One way to limit the voter's choices to just two candidates is to permit two, and only two, political parties. This solution is obviously unpalatable, and is probably unconstitutional. Nevertheless, it might be useful to consider why this would be objectionable.

A typical party primary allows voters to choose among candidates preselected by the political parties. It is entirely possible that a candidate might be supported by the *voters* but, for whatever reason, not by the party itself; such a candidate could never win the nomination, and (were the political parties the only way to be on the general ballot) could never run for office.

One way around this emerged in 1936, when Washington State introduced a new method of choosing candidates known as a *blanket primary*. In a blanket primary, the candidates self-declare their political affiliation; the candidate declaring for a political party who wins the most votes *becomes* that party's candidate in the general election. It is entirely possible for a party's candidate to be someone supported by the electorate and not by the party itself. Consequently, *all* political parties object to blanket primaries. However, the Washington State Supreme Court ruled, in *Anderson v. Millikin* (1935), that the blanket primary system was constitutional. The U.S. Supreme Court declined to hear the case, and the blanket primary spread to Alaska in 1947.

That could have been as far as it went: between them, Washington and Alaska elected just 10 congressmen. But in 1996, California voters passed Proposition 186, instituting a blanket primary system in California, with its 52 congressional seats. The Democratic Party filed suit, and in *Democratic Party of California v. Jones* (2000), argued that Proposition 186 was a violation of their 1st Amendment rights to free association. In particular, by forcing them to support a political candidate selected by the voters, members of the political party were forced to associate with a candidate they might not have chosen. The Supreme Court agreed in a 7 to 2 decision, effectively invalidating the primary system in three states.

However, the state of Washington, having used a blanket primary for 70 years, chose to reinvent it. The legislature passed Initiative 872, which established a *nonpartisan blanket primary*. On the surface, this is similar to a blanket primary: voters choose from a list of candidates to decide who will run in the general election. But there is a crucial difference: the primary does *not* choose the party nominees. Rather, the top two vote-getters are the candidates, regardless of their declared affiliation.

Once again, this system allows the voters, and not the political parties, to determine the candidates. And once again, the political parties launched a lawsuit to repeal it. This time, the Republican Party argued the case before the Supreme Court. However, because the nonpartisan blanket primary is designed to limit the candidates who appear on the ballot, and not determine which candidate receives a party's nomination, on March 18, 2008, the Supreme Court declared it constitutional, in *Washington Grange v. Washington State Republican Party* (2008). Two years later, California voters passed Proposition 18, which instituted a similar system.

Runoff Voting

Another way to decide which two candidates are allowed to run involves abandoning the primary system altogether. Louisiana did this in 1978, using what is usually called a *Cajun primary*. Under this system, anyone running for public office registers with the state board of election, and is placed on the ballot. As in plurality systems, voters cast their vote for just one candidate. However, in order to win the election, a candidate must win a majority of the votes cast.

In 2012, voters in Louisiana's 2nd congressional district, which includes most of the city of New Orleans, had a choice among two Democrats, two Republicans, and one Libertarian. Cedric Richmond, the incumbent, won 55.2% of the vote, and was elected.

Obviously, Richmond would have won a two-way race against any of the other candidates, so he was the Condorcet winner. In general, this system satisfies the majoritarian principle—as long as some candidate wins a majority. But what happens when no candidate has the majority? In these cases, a second runoff election is held a few weeks later between the two top vote-getters. This occurred in Louisiana's 3rd congressional district, where three Republicans, a Democrat, and a Libertarian appeared on the 2012 ballot. No candidate won a majority, so on December 8, the two top vote-getters, Jeff Landry and Charles Boustany, Jr. (both Republicans) faced each other. Boustany won with 60.9% of the vote.

The Cajun primary is more widely known as *majority runoff voting*, and is used extensively in France. (Although many laws and legal institutions in Louisiana trace their ancestry to French law, the Cajun primary is an original creation.) Meanwhile, since the systems in California and Washington present voters with a choice of two and only two candidates, they are known as *top-two* systems, though they are also versions of majority runoff voting.

Opponents raise a number of objections to such systems, one of the more common being that it's possible that a political party might have no candidates on the final ballot. But the Supreme Court has repeatedly held that political parties do not have constitutional rights, so preserving the power of a political party is not constitutionally relevant. The systems must be assessed in terms of how they affect the voters and the choices made by them.

A more serious objection to majority runoff systems is that, unless a candidate wins a majority, a second election must be organized, which means candidates must campaign twice, and voters must go to the polls twice. But this is true in a traditional primary system as well. In fact, as events in Louisiana's 2nd district showed, a candidate might only need to run one campaign, and voters may only have to go to the polls once. Indeed, only one of the six congressional elections held in Louisiana in 2012 required a runoff; the other five were decided by the first election.

Another problem with runoff systems may be illustrated by the results in Louisiana's 3rd district. The two candidates who advanced to the runoff, Charles Boustany and Jeff Landry, won 45% and 30% of the vote, respectively. Democrat Ron Richard came in a somewhat distant third, with 22% of the vote, but it's easy enough to imagine a situation where three (or more) candidates had very nearly equal shares of the vote: for example, in a five-candidate race, where the top four candidates won 21%, 20%, 19.9%, and 19.8% of the vote, how reasonable is it to dismiss the third- and fourth-place candidates on the basis of a 0.1% difference in their vote shares?

One way to resolve this issue is to treat the space on a runoff ballot as an apportionment problem. In an apportionment problem, states "buy" congressmen with their share of the national population. In the same way, candidates could "buy" spaces on a runoff ballot with their share of the vote. For example, the Jefferson and Webster methods of apportionment choose a district size; divide each state's population by the district size to find a quota; then round down (Jefferson) or to the nearest whole number (Webster) to determine the number of congressmen due each state.

A variation, applicable to the runoff problem, is to choose the percentage of vote necessary to win a space on the runoff ballot; divide each candidate's share by that percentage; then round down (following Jefferson) or to the nearest whole number (following Webster). Candidates who win at least one space are on the runoff ballot; candidates who fail to win at least one space are not.

When applying the Jefferson or Webster method to the apportionment problem, we had to solve the difficult problem of finding a divisor that gave us the right number of seats. But when applying this approach to the runoff problem, we can take advantage of the fact that candidates are permitted only have one space on the runoff ballot, so we can simply choose a percentage that gives the leading candidate only one space. To give the voters as many choices as possible, we might use the *lowest* percentage that gives the leading candidate one space.

For example, consider an election with four candidates x, y, z, and w, where the percentage of the vote won by the candidates is 40%, 24%, 21%, and 15%. Applying Jefferson's method, which rounds down, we note that a divisor of 20% would give x a quota of 40% ÷ 20% = 2 seats, so we can use any divisor greater than 20%: thus, we might use 21%, which gives the candidates quotas of 1.905, 1.14, 1.05, and 0.67 respectively, which become 1, 1, 1, and 0 after rounding down; thus x, y, and z win a place on the runoff ballot. With Webster's method, a divisor of 26% would give x a quota of 40% ÷ 26% = 1.54, which would round up to 2, so we use a divisor of 27% and obtain quotas of 1.48, 0.89, 0.81, and 0.52, which are rounded to 1, 1, 1, and 1, giving all four candidates a place on the runoff ballot.

While there's nothing wrong with giving all candidates a place on the runoff ballot, it would seem to defeat the purpose of *having* a runoff election. In fact, any divisor method can lead to such a scenario. The Jefferson method, applied to an election where the candidates had vote shares of 30%, 28%, 26%, and 16%, would also give all candidates a slot on the runoff ballot.

A more objectionable feature is that it's possible for a candidate who wins a *majority* of the vote to face another candidate in a runoff ballot. For example, if the vote shares of the four candidates were 60%, 35%, 3%, and 2%, the first two candidates would (under either Webster or Jefferson) be set against each other in a runoff election. While we could draft a law that only requires runoff elections when no candidate has a majority, the fact that we must write this in as a special case is both legally and mathematically inelegant. It would be much better if we could draft a single rule that fits all cases.

One solution is the *Next Two Rule*, proposed in 2011 by Steven Brams of New York University and D. Marc Kilgour of Willfrid Laurier University in Waterloo, Canada. This rule constructs the runoff ballot by adding candidates as follows. First, the candidate with the most votes is automatically put on the ballot. Next, if a candidate's share of the vote, plus the share of the vote of the next-highest vote-getter, exceeds the share of the vote by any candidate already on the runoff ballot, that candidate is also added to the ballot.

For example, consider our first election, with the four candidates x, y, z, and w receiving 40%, 24%, 21%, and 15% of the vote. Candidate x, with the highest percentage of the vote (40%), is automatically on the runoff ballot; y's share of the vote (24%), together with candidate z's share (21%), is equal to 45%. Since this is greater than x's share, y also wins a place on the ballot.

Next, z's share (21%), together with w's share (15%), is 36%. Since this is greater than the share of another candidate on the runoff ballot (namely y), z also wins a place on the runoff ballot. On the other hand, w has no next-highest vote getter; in general, the last place vote-getter will always be dropped, so the Next Two rule guarantees that a runoff election will have fewer candidates. And if any candidate wins a majority, the second-place candidate cannot win a place on the runoff ballot (since it would be impossible for the vote shares of the second- and third-place candidates to be greater than the share won by the first-place candidate): thus majority winners need not face others in a runoff.

Instant Runoff Voting

We can avoid the runoff election entirely through the simple expedient of having voters submit their preference lists. For example, consider a Cajun primary system applied to the election profile in table 5. As we saw earlier, if each voter only cast a ballot for their first-place candidate, x would win 150 votes; y would win 140 votes; z would win 130 votes; and w would win 100 votes, so no candidate has a majority. Since x and y have the most votes, they would face each other in a runoff.

Because voters can only vote for one candidate, and (under our current assumptions) they will only vote for their favorite candidate, it follows that the 150 voters with ranking $x > w > y > z$ would still vote for x, and the 140 voters with ranking $y > w > z > x$ would still vote for y.

But consider the 130 voters with ranking $z > y > w > x$. Since they must choose between x and y, these voters would submit a ballot for y, their preferred candidate. Likewise, the 100 voters with ranking $w > z > x > y$ would submit a ballot for x. In the end, x would receive $150 + 100 = 250$ votes, while y would receive $140 + 130 = 270$ votes, and y would be the winner of the runoff election.

In fact, if we have the preference lists and the election profile, we need not eliminate all but two of the candidates immediately: we could determine the outcome of any election from the profile. In *instant runoff voting* (IRV), the candidate with the fewest first-place votes is eliminated, with the election profile rescored to reflect the absence of the eliminated candidate. We could actually apply any runoff mechanism we want: for example, we could apply the Next Two rule to determine which candidates to eliminate. IRV is used in municipal elections in San Francisco, Minneapolis, and Portland (Maine), as well as elections to Australia's lower house of parliament and Irish presidential elections.

In 2006, Burlington, Vermont instituted instant runoff voting for its mayoral elections. Bob Kiss, a Progressive, won the election that year; fortunately, for the credibility of instant runoff voting, Kiss would have won under a plurality vote, so the outcome of the election was consistent with the expectations of less sophisticated voters.

In 2009, Kiss ran for re-election against Kurt Wright, a Republican, and Andy Montroll, a Democrat. There are six ways that voters could rank the three candidates (which we'll designate K, W, and M); in addition, some voters only voted for one of the candidates (lending some weight to the claim that alternative methods of voting are too confusing for voters). The numbers of voters who submitted each of the nine submitted rankings were:

M > K > W, 1332
M > W > K, 767
K > M > W, 2043
K > W > M, 371
W > M > K, 1513
W > K > M, 495
W only, 1289
K only, 568
M only, 455

Some write-in candidates also received votes.

Consequently, counting the votes for just one candidate as giving that candidate a first-place vote, Montroll received 2,554 first-place votes; Kiss 2,982; Wright, 3,297; and various write-in candidates, some small number of votes. No candidate had a majority, invoking the instant runoff mechanism. The write-in candidates would have been eliminated almost immediately, but eventually the candidate list would have been winnowed to Montroll, Kiss, and Wright.

Since Montroll had the fewest first-place votes, he was eliminated and the ballots re-ranked, so the 1,332 voters with preference list M > K > W would become 1,332 voters with preference list K > W. They would be joined by 2,043 voters with preference list K > M > W. The votes for Montroll only would be eliminated entirely, as they are effectively invalid votes. The new election profile would be:

K > W, 3746
W > K, 2775
W only, 1289
K only, 568

This gave Kiss $3,746 + 568 = 4,314$ votes and Wright $2,775 + 1,289 = 4,064$. Since Kiss had a majority of the valid votes cast, Kiss won the election. And, since there were only three candidates, the Next Two rule would have produced identical results.

However, a close analysis of the tallies yields some disturbing observations. First, Wright, the Republican candidate, actually won the plurality vote. With instant runoff voting so new to the electorate, the fact that the person who would have won a traditional election lost under the new mechanism struck many as a defect; certainly, a losing candidate could make much of it, and Wright's supporters launched a suit on this basis. However, the lawsuit was dismissed and the election results upheld: the Constitution does not require a specific system of voting.

More troublesome was the fact that Montroll was the Condorcet winner. The fact that neither the plurality winner nor the Condorcet winner won the mayoral election was enough to persuade voters to repeal instant runoff voting following the 2009 election.

Since there were only three candidates, instant runoff voting and majority runoff voting would have yielded the same results. Instant runoff voting, as well as the Cajun primary and the nonpartisan blanket primary, can violate the Condorcet winning condition. This has been presented as a serious flaw in these systems, but the flaw is literally more apparent than real in the latter for several reasons. Because voters only vote for one candidate in the Cajun primary and the nonpartisan blanket primary, it is generally impossible to identify the Condorcet winner. Only instant runoff voting collects enough information to identify a Condorcet winner. If a candidate wins a majority, the candidate is also the Condorcet winner; however, the reverse is not always true. For example, in an election profile where voters split their first-place votes equally among candidates a, b, and c, but all voters rank d in second place, d will be the Condorcet winner. But if voters only list their first-place votes, as they do in the nonpartisan blanket primary and the Cajun primary, *d* would receive 0 votes.

Should we conclude that ignorance is bliss, and argue that it is better *not* to collect the preference lists? Not at all! The simple and obvious solution is to collect the preference lists and use them. If a Condorcet winner exists, declare that person the winner; otherwise, invoke the runoff mechanism (possibly beginning with the elimination of the Condorcet loser, if such a candidate exists). The inclusion of this simple provision might have saved instant runoff voting in Burlington, Vermont, and should be a lesson to those who wish to change the voting system.

Single-Peaked Preferences in Abstract Policy Spaces

The possibility that the winner of an election might not be the Condorcet winner, and might even be the Condorcet loser, suggests a way of evaluating the different voting systems. We can do this by simulating elections and seeing how often a voting rule fails to select the Condorcet winner, or declares victory to the Condorcet loser. A standard approach is to use a *spatial model* of voters and candidates.

The basic concepts involved are straightforward. First, the *policy space* represents an abstraction of all the components of a person's preferences. For example, the left/right political spectrum is an example of a one-dimensional policy space. Higher dimensional policy spaces are possible: the only requirement is that there be some sense of "more extreme" or "less extreme" along each of the various directions.

The policy space applies to both the voters and the choices available to them. For illustrative purposes, imagine the voters (A through L) and candidates (*w*, *x*, *y*, and *z*) spread out along a one-dimensional policy space as shown in figure 1. If we assume that voter preferences are *single-peaked*, then the farther away a candidate is from a voter's position, the lower the candidate appears on their preference list. If we also assume that voter preferences are *symmetric*, then only the distance matters, not the direction.

Figure 1. Voters in a One-Dimensional Policy Space

Both of these assumptions are reasonable. First, the assumption that voter preferences are single-peaked translates into the idea that if a candidate is too radical for a voter, candidates even more extreme are even less appealing. The single-peaked assumption seems reasonably reflective of how people make choices: at no point does "even worse" become "better."

The symmetry assumption is a little more dubious, since one might expect voters on the left to prefer candidates slightly to the left of their position over those who are slightly to the right. Nevertheless, theorists generally accept the symmetry assumption for three reasons. First, it makes the analysis much easier. Second, failure of symmetry would translate into a voter favoring those on the extremes over those very near to their own position but on the "wrong side," so that a slightly left-of-center voter would prefer extreme leftists to centrists. Third, and perhaps most important, theorists who incorporate asymmetric preferences obtain essentially the same results as theorists who assume symmetry.

These assumptions allow us to recover the preference list for any voter by listing the candidates in order of increasing distance from the voter. So, for example, voter C will have preference list $x > w > y > z$. Since we can do this for every voter, we can construct the election profile; and once we have the election profile, we can determine the election outcome under any voting rule. Under plurality rule, then, voters vote for the candidate closest to them in the policy space. For the distribution of voters and candidates in figure 1, x would win the election, since x is closer to more voters than any other candidate.

The alert reader might observe that the one-dimensional policy space prohibits certain preference lists: for example, it is impossible to locate a voter whose preferences are $y > z > w > x$. This impossibility is primarily because we have limited ourselves to one-dimensional policy spaces; if we allow for multidimensional policy spaces, we can support any preference listing.

To simulate an election, Keith Dougherty, in the political science department at the University of Georgia, and Julian Edward, in the mathematics department at Florida International University, assumed voters and candidates were distributed in a one-dimensional policy space in a variety of ways. Not all election profiles yield a Condorcet winner, but suppose we focus on those that do. Dougherty and Edwards found that plurality voting will *fail* to select the Condorcet winner about 30% of the time. Even worse, if a Condorcet *loser* exists, plurality selects the Condorcet loser about 20% of the time. What about majority runoff, instant runoff voting, or the Borda count? These systems never select the Condorcet loser, yet they still fail to select the Condorcet winner about 15% of the time.

How Bad Is Plurality?

However flawed we might believe plurality systems to be, they have the advantages of familiarity, simplicity, and tradition. Consequently, any proposal to replace plurality with another system will require considerable evidence that the new system is superior. Moreover, if the new system produces objectionable results, it will be short-lived: Burlington, Vermont abandoned instant runoff voting after it failed to elect the plurality and the Condorcet winners, and the future of majority runoff in California may be threatened by the prevalence of districts where voters had to choose between two candidates from the same party.

One approach might be to identify what we want in a voting rule. If we could find a voting rule that meets *all* of our requirements, it follows that any system that fails to meet *any* of the requirements is not as good, and no rational person would choose the second system over the first. For a variety of reasons, it is easier to talk about social welfare function instead of voting rules, so we will focus on them.

There are a number of features we might want to include in our social welfare function, but three seem obvious: *non-imposition*, which means that for any candidate, there is some election profile that puts that candidate at the top of the ranking; *neutrality*, which means that the names of the candidates don't matter; and *nondictatorship*, which means that no voter can single-handedly determine the outcome of the social welfare function, regardless of what the other voters do.

There are two other features that seem innocuous enough. These can be viewed in the following context. Imagine that, based on pre-election polls, a prediction is made that the social welfare function will rank $x > y$. This does not mean that x will win the election, though it does mean that x will do better than y. Imagine this situation from the viewpoint of y's supporters. Since x will do better than y, they must campaign to alter the election profile to have any chance of winning. But suppose it is impossible to alter any voter's relative ranking of x and y. If a voter ranked $x > y$, then nothing will persuade them to rank $y > x$. This situation might well cause dismay among y's partisans, as it appears there is no way to alter the election results.

But a campaign worker (perhaps one who has read this book) makes the following claim: it *is* possible for y to win *without* a single voter changing their relative ranking of x and y. Suppose our voting system is plurality, and we have the following election profile:

$x > y > z$, 150 voters
$y > z > x$, 125 voters
$z > y > x$, 50 voters

Plurality rule would return a societal ranking of $x > y > z$, so x would win the election.

Consider the 50 voters with preference list $z > y > x$. The preference $z > y > x$ can be regarded as how they truly feel about the candidates, and if they submit a *sincere ballot* that reflects their true feelings, they will cast their vote for z, and x will win—from their point of view, the worst outcome. On the other hand, they might submit an *insincere ballot*. Suppose they feign a preference $y > z > x$. Under plurality rule, y would receive their votes, and in the election, y would win $125 + 50 = 175$ votes, and emerge the victor. By *strategic voting*, these voters have achieved a better outcome (the victory of their second choice instead of their last choice). This goes by the unfortunate name of *manipulation*, which brings to mind images of collusions in smoke-filled back rooms, though in mathematical politics it's a purely neutral term: it merely means that a voter can obtain a better outcome by submitting an insincere ballot.

This type of strategic voting (or lack thereof) is the basis of claims that third-party candidates like Roosevelt or Nader led to the defeat of a major party candidate. However reasonable these claims might be, it's worth remembering that in plurality voting, we do not collect enough information to prove these allegations. While it is reasonable to suppose that some of Nader's supporters would have voted for Gore in 2000, it's at least conceivable they might have voted for Bush or abstained entirely.

A more serious problem with strategic voting is that voters support candidates who can win, and not necessarily candidates they prefer. As a result, elected officials may receive a flawed view of how the electorate really feels, claiming an electoral mandate or political capital where none exists. We might insist that the ideal voting system satisfy *independence of irrelevant alternatives*. In particular: the social welfare function can alter the relative ranking of x and y *only* if at least one voter changes their relative ranking of x and y. In the earlier case, the relative ranking of x and y did not change; only the ranking of z (the irrelevant alternative) changed, and that was enough to change the outcome. Hence plurality fails independence of irrelevant alternatives.

As it turns out, *all* of our methods fail independence of irrelevant alternatives. This is easiest to illustrate with a Borda count. Suppose $x > y$ in a Borda count. Consider voters who ranked $z > y > x$. If they submitted $y > z > x$, then y's point total would increase (since they have more first-place votes), possibly allowing them to win without anyone changing their relative ranking of x and y. Again, these voters have improved the outcome by strategic voting and insincere ballots.

There is a more bizarre way to alter election outcomes. Consider an instant runoff system applied to an election with three candidates x, y, and z, and 500 voters. Suppose the election profile at the beginning of election day is:

$x > y > z$, 125 voters
$y > z > x$, 150 voters
$z > x > y$, 175 voters
$x > z > y$, 50 voters

Since no candidate has a majority, the instant runoff mechanism would be invoked. Based on this profile, y will have the fewest first-place votes and be eliminated. The ballots showing preference $x > y > z$ would be re-ranked $x > z$; the ballots showing preference $y > z > x$ would be re-ranked $z > x$, and so on. Our new profile will be:

$x > z$, $125 + 50 = 175$ votes
$z > x$, $150 + 175 = 325$ votes

therefore z wins the election.

Of course, z's supporters don't know they will win the election, so they might continue to campaign. Imagine that last-minute campaigning persuaded the voters who started the day with preference list $x > z > y$ to alter their preference list to $z > x > y$. Now z has even more first-place votes than before. But our profile is now:

$x > y > z$, 125 voters
$y > z > x$, 150 voters
$z > x > y$, 175 voters
$z > x > y$, 50 voters

Now x has the fewest first-place votes, so they will be eliminated. After re-ranking to reflect x's elimination, the new profile is:

$y > z$, $125 + 150 = 275$ votes
$z > y$, $175 + 50 = 225$ votes

so now y wins the election!

The possibility that raising the rank of a winning candidate (z) can cause them to lose the election seems objectionable. This leads to the desirability of *monotonicity*, which is equivalent to requiring that a winning candidate will still win if more voters support him or her. When a monotonicity violation occurs, additional support for a candidate causes them to lose. In practical terms, a monotonicity violation would occur whenever a group of voters supported a losing candidate who would have won *without* their support.

Since majority runoff and instant runoff voting produce identical results when there are three candidates, the preceding shows both systems fail monotonicity. In contrast, plurality satisfies monotonicity: more votes for the winning candidate provide a larger margin of victory. The Borda count likewise satisfies monotonicity, since raising the rank of a winning candidate will raise their point total at the expense of another candidate. Note that we are assuming that a voter *only* increases the rank of the winning candidate. If a voter increases the rank of a winning candidate *and also* increases the rank of a nonwinning candidate, it's possible for the latter to gain enough points to win. However, this does not constitute a monotonicity violation, since we have altered the ranking of another candidate as well as that of the winning candidate.

We have five conditions: neutrality, non-imposition, nondictatorship, independence of irrelevant alternatives, and monotonicity. All of the voting rules thus far discussed fail independence of irrelevant alternatives and can violate the Condorcet winning criterion. Majority runoff and instant runoff also fail monotonicity. Since the Borda count and plurality satisfy monotonicity, we might regard them as superior to majority runoff and instant runoff. The Borda count is more complicated, since voters must actually rank candidates. This leaves plurality as a simple system that satisfies more of our requirements than any other system.

Arrow's Theorem

But perhaps we just haven't been sufficiently imaginative. Could a sufficiently clever person invent a system of voting that satisfies all of these criteria? Unfortunately, the answer is *no*.

In 1950, Kenneth J. Arrow published a paper modestly titled "A Difficulty in the Concept of Social Welfare." In this paper, Arrow, who graduated from City College in New York in 1940 with a degree in mathematics and social science before pursuing graduate work in economics, proved a stunning result: If there are more than two candidates, *no* social welfare function satisfies all five requirements. This result led to Arrow being awarded the Nobel Prize in Economics in 1972.

For a result that led to a Nobel Prize, Arrow's theorem is remarkably simple, and the basics of the proof are easily understood. This should not detract from its importance: indeed, one definition of genius is seeing the obvious thing that everyone else has ignored. Arrow actually proved that any social welfare function that satisfied independence of irrelevant alternatives, monotonicity, and non-imposition would fail nondictatorship. For convenience, we'll add neutrality and assume the social welfare function is resolute: in other words, the names of the candidates don't matter, and ties are resolved in some nonrandom fashion. Arrow's theorem still holds even for nonresolute and non-neutral systems, but the proof is more complicated.

Suppose we have a resolute social welfare function that satisfies independence of irrelevant alternatives, monotonicity, non-imposition, and neutrality. Let there be at least three candidates a, b, and c. Non-imposition means that any candidate *can* win, so let us begin with an election profile where a wins. This means the social welfare function ranks $a > b$ (since winners must outrank nonwinners).

Imagine putting all of the voters into one of two rooms. Room A contains all voters whose preference lists indicated $a > b$, regardless of how they ranked c (thus, the room would include voters with preference list $a > b > c$, $a > c > b$, and $c > a > b$), while Room B includes all those whose preference lists indicated $b > a$, again regardless of how they ranked c (so this room would contain voters with preference list $b > a > c$, $b > c > a$, and $c > b > a$).

Now let all the voters submit new ballots as follows. First, let some of the voters in A submit a preference a > b > c; we send these voters to one corner of the room, which we will call A1. Let the remaining voters in A submit a preference c > a > b, and send them into a different corner of the room, which we will call A2. At the same time, let all voters in B submit new preferences b > c > a. Note that all voters in A, whether they are in corner A1 or A2, still rank a > b, and all voters in B still rank b > a. Thus no voter has changed their relative ranking of a and b. Because we assumed our social welfare function satisfies independence of irrelevant alternatives, and our original profile returned a > b, then our social welfare function still returns a societal ranking of a > b (see figure 2). But where is c?

Because the social welfare function doesn't produce ties, there are only two possibilities: a > c, or c > a. First, we might have a > c. Note that *only* the voters in A1 ranked a higher than c, while all other voters (those in A2 and in B) ranked c > a. By neutrality, monotonicity, and independence of irrelevant alternatives, this means that if the voters in A1 rank *any* candidate x first, then x will win the election. For obvious reasons, the voters in A1 are said to form a *dictating set*.

On the other hand, we might have c > a (since voters might have changed their relative ranking of a, c, it's possible that c is now the winner). Since the social welfare function still returns a > b, then c > a and a > b means (by transitivity) c > b. But only the voters in A2 ranked c higher than b. As before, this means that if this group ranks *any* candidate x first, then x will win the election: thus A2 is a dictating set.

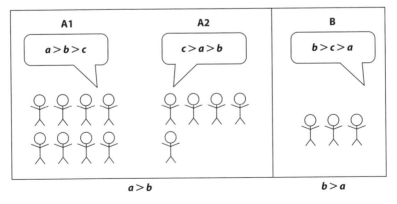

Figure 2. If a > c, then A2 voters must go to room B

Either the voters in corner A1 or the voters in corner A2 form a dictating set. Now, let the dictating set stay in room A, and send the rest of the voters into room B. This time, change everyone's preferences to match their room, so that everyone in room A ranks $a > b$, and everyone in room B ranks $b > a$. Because the voters in room A are a dictating set, the voting system returns $a > b$. This means we can repeat the process and find a smaller group within room A that forms a dictating set. Since, at every stage, some voters are sent from room A into room B, but voters are never returned from B back to A, it follows that we will gradually reduce the number of voters in room A. The process ends when we have just one voter in room A: the dictator.

Arrow's theorem is a statement about an unavoidable *logical* consequence of our assumptions: no matter how clever we are, we can never create a voting system that satisfies monotonicity, independence of irrelevant alternatives, neutrality, non-imposition, *and* nondictatorship. Just as there is no point in looking for a triangle with two right angles, there is no point in trying to invent a voting system that satisfies all five conditions.

Breaking Arrow

The usual interpretation of Arrow's theorem is that a "perfect" voting system is impossible. However, this is an oversimplification. First, Arrow's theorem requires that there be at least three candidates, and that voters be able to adopt any preference listing. If there are only two choices (for example, a yes/no vote), plurality is the only voting rule that satisfies all five conditions (a result known as *May's theorem*). Plurality also satisfies all five conditions in a one-dimensional policy space, because certain preference listings are impossible. Studied from another viewpoint, if voters could *only* choose between two candidates, or if voters were sufficiently simple-minded and judged candidates *only* on the basis of where they resided in the left/right political spectrum, plurality would be an ideal form of voting.[4]

Of course, voters usually have a choice between three or more candidates, and we hope most voters are *not* so simple-minded as to rank candidates solely on the basis of the left/right political spectrum. Even so, we can avoid

---

4. Likewise, we *can* find triangles with two right angles, as long as we are willing to abandon plane (Euclidean) geometry and consider the geometry on other surfaces.

Arrow's theorem, provided we allow for a somewhat radical departure from traditional social choice theory.

Traditional social choice theory requires that voters have a ranking of all possible choices: thus the Borda count actually relies on voters ranking the candidates. But consider ordering from a restaurant menu. If diners actually ranked all items on the menu, they would have no problem ordering even after they were told their first, second, and third choices were unavailable. However, experience shows that identifying your fourth choice item *is* difficult; in such cases, we're more likely to fall back on some standard item that we know we'll enjoy, as opposed to searching through the menu for our fourth choice.

As an alternative, consider *grading* the choices. For example, we might assign a grade of excellent, good, acceptable, poor, or reject. There are three advantages to grading. First, it is easier for the voters: ranking requires considering how a candidate rates relative to all others; grading only requires considering a candidate against some internal measure. Next, voters have a better ability to express their true feelings. If candidates are merely ranked, two voters with preference $x > y > z$ have identical effects on the outcome. But one voter might rank x, y, and z as excellent, good, and acceptable, while the other might rank them as good, poor, and reject; this could alter the outcome to better reflect voter sentiment. This last feature leads to the key virtue of grading: Arrow's theorem relies on the ranking of candidates; if the candidates are not ranked, Arrow's theorem does not apply.

For these and other reasons, Michel Balinski and Rida Laraki, of the École Polytechnique in France, have proposed using *majority grading* as a better means of deciding electoral outcomes. The majority grade is found as follows. As with other social choice functions, we begin with a preference list, though one based on grades and not ranks. For example, suppose there are three candidates and seven voters who submit as shown in table 6.

What shall we do with the ballots? One possibility is to assign a reject grade 0 points; poor, 1 point; acceptable, 2 points; good, 3 points; and excellent, 4 points; then compute the mean: in effect, we are computing the candidate's grade point average!

One objectionable feature of the mean is that it's subject to distortion by extreme values. This is most evident in situations where there is no upper limit. For example, if there are ten homes for sale, nine of which are priced at $200,000 each and the tenth priced at $5,000,000, then the mean home price

Table 6.   Grading the Candidates

Voter	Candidate x	Candidate y	Candidate z
1	excellent	good	reject
2	good	good	poor
3	good	poor	excellent
4	good	reject	excellent
5	reject	acceptable	excellent
6	reject	reject	acceptable
7	reject	reject	acceptable

will be \$680,000, which gives a distorted view of the typical home price. Thus, a better approach is to use the *median*: the middle value, when all the values are put in order from least to greatest. In this case, the median is \$200,000, which gives a much better view of the typical home prices. If considering an odd number of values, there's a unique middle value that we take as the median. If there's an even number of values, as in this case, there are two middle values, so we take the value midway between them, or create one (in the case where our values are ranks, we might say "excellent minus" is midway between excellent and fair).

Because the median is less subject to such distortion, Balinski and Laraki (among others) suggest using the median. In this case, the grades for our candidates (in order, with the median **in bold**) will be:

x: excellent, good, good, **good**, reject, reject, reject

y: good, good, acceptable, **poor**, reject, reject, reject

z: excellent, excellent, excellent, **acceptable,** acceptable, reject, poor

Thus x wins, with a majority grade of good.

Majority grading avoids Arrow's Impossibility Theorem. Moreover, it has one unique feature, compared to the other systems of voting we've examined: it also satisfies independence of irrelevant alternatives. This is because a candidate's median grade can *never* be altered by changing another candidate's grade. As a result, the system is non-manipulable: voters cannot obtain a better outcome by submitting an insincere ballot.

However, there are some troublesome features of majority grading. One in particular is the one known as the *no-show paradox*. Suppose that in the election, only voters 1 through 5 actually cast ballots. With just these five voters, the grades, again with the median in bold, would be:

x: excellent, good, **good**, good, reject
y: good, good, **acceptable**, poor, reject
z: excellent, excellent, **excellent**, poor, reject

Thus z wins if voters 6 and 7 are "no-show voters." What makes this a paradox is that voters 6 and 7 rank z *above* candidates A and B.

We can view the no-show paradox as a variation of a monotonicity violation, but there is an even more serious situation where the paradox will emerge. In most U.S. elections, voting results are reported by precincts. If we imagine that voters 1 through 5 are in one precinct, while voters 6 and 7 are in another, we would have a situation where z wins both precincts but loses the general election! This is a version of *Simpson's paradox*, which we'll discuss later in the book. A similar situation would occur if we used the mean grade instead of the median.

Approval Voting

The precincting version of the no-show paradox should be viewed as a rather serious flaw of majority grading. Moreover, there's no guarantee that majority grading won't fall victim to some result analogous to Arrow's Impossibility Theorem. Thus, while majority grade has considerable potential as an alternative way to elect a candidate, it's worth reconsidering the implications of Arrow's theorem. Arrow himself considered his result to be a *possibility* theorem. In particular, although we can't have neutrality, non-imposition, nondictatorship, independence of irrelevant alternatives, *and* monotonicity, we can have any four of them.

Since it seems objectionable to abandon non-imposition, neutrality, or nondictatorship, we are left with a choice between monotonicity and independence of irrelevant alternatives. As we saw, none of our systems satisfy independence of irrelevant alternatives; majority runoff and instant runoff also fail monotonicity. Since both plurality and the Borda count satisfy four of the five requirements, it seems that they are superior to both majority runoff and instant runoff voting. They also avoid the no-show paradox. Plurality is also simpler and more familiar, adding to its appeal as a form of voting. The major flaw with plurality is that it violates both the Condorcet winning condition and the Condorcet losing condition.

If we are willing to accept a voting system that fails these two, there is another voting system worth considering: approval voting, which was invented in 1978 by Guy Ottewell; and also, independently, by Robert Weber; and promoted by political scientist Steven Brams and mathematician Peter Fishburn. Under an approval system, voters can cast a vote for *all* candidates they approve of; the candidate with the most approval votes wins the election. If there are six candidates, a voter could cast a vote for any number of them (including all six, though that would simply give *all* candidates one more vote).

It's convenient to think of approval voting as follows. Let every voter have a preference list, and a threshold of acceptability. If a candidate is above the threshold, the voter indicates approval of that candidate; otherwise, the voter omits that candidate. Voters with a high threshold approve of very few candidates; voters with a low threshold approve of most.[5]

While approval voting violates the majoritarian principle, Condorcet winning criterion, Condorcet losing condition, monotonicity, and independence of irrelevant alternatives, there are reasons to consider it superior to any of the other voting rules we have discussed. The reason emerges from *how* it fails these things.

First, consider the majoritarian principle and the Condorcet criterion. Suppose there are three candidates x, y, z, and suppose there are five voters with preference listings as shown, followed by the candidates they approve of:

Voter 1: $y > z > x$, approves x, y, z
Voter 2: $y > z > x$, approves y
Voter 3: $z > x > y$, approves z, x
Voter 4: $x > y > z$, approves x
Voter 5: $y > x > z$, approves y, x

Because the voters have different thresholds, they will approve of different portions of their list. For example, voters 1 and 2 have the same ranking of the candidates, but voter 1 is very forgiving and approves of all three candidates, while voter 2 is more selective and only approves of y.

The tally of approval votes is: x, 4 votes; y, 3 votes; and z, 2 votes. What's important to realize is that the election results *only* indicate the number of

---

5. One could view majority grade as a generalization of approval voting, or approval voting as a special case of majority grade. However, not everything that applies to approval voting applies to majority grade, and vice versa, so the two systems deserve separate treatment.

approval votes; the voter preferences are concealed. This is an important feature, because while approval voting can violate the majoritarian principle, the Condorcet winning condition, the Condorcet losing condition, *and* monotonicity, the manner in which these violations occur seems to be minimally objectionable.

To begin with, consider the majoritarian winner. Because voters can approve of two or more candidates, it's possible for more than one candidate to win a majority of votes: indeed, in this example, both x and y won a majority of votes, though x won a larger majority and the approval vote. The actual majoritarian winner, as determined by which candidate would defeat any other candidate in a two-way race, is y.

Consider trying to argue that a majoritarian violation occurred. This would require arguing that x, the candidate who won more approval votes than y, would have lost to y in a two-way race. In a plurality system, we could argue as follows: While x won 4 votes, y won 3, and z won 2, it's possible that those who voted for z would have preferred y, so in a two-way race between x and y, y would have won $3 + 2 = 5$ votes to x's 4 votes.

Such an argument carries considerable weight in plurality voting: it is essentially the argument that third-party candidates are "spoilers" in presidential elections. But the argument is considerably weaker in an approval system. The argument would only apply to voters whose preferences are $z > y > x$, and if such voters truly preferred y over x, they would be able to approve of both z and y. Their failure to approve of y as well as z reflects the fact that, as far as these voters are concerned, x and y are equally unacceptable.

A similar situation occurs when trying to argue that a violation of the Condorcet winning condition has occurred (which it has, in this case): one would have to argue that the ballots indicating approval of two or more candidates "really" meant that the voter only approved of a specific candidate.

Approval voting can also fail monotonicity. Suppose a candidate w wins under a given election profile. A monotonicity violation could occur as follows. Consider a voter who originally approved of w. This voter might simultaneously raise w's ranking and the threshold of acceptability, so that his or her ballot no longer indicates approval of w. Such a situation isn't too unrealistic: consider a scandal that affects many candidates, but w less than the others. The voter might raise w's ranking relative to the others while at the same time disapproving of w. Thus w would end with fewer votes and a potential electoral loss.

On the other hand, suppose w fell short of some voter's threshold of acceptability. The voter might raise w's ranking and *lower* his or her threshold of acceptability (but still leaving it above w's ranking). This would give one more approval vote to candidates other than w and potentially cause w to lose the election. Remember that we have assumed our system resolute. If w wins by one vote, but then loses one vote and ties with the second-place candidate, the system might resolve ties against w, causing them to lose.

In the first case, a voter who originally approved of w now fails to approve of w; and in the second, a voter who did not approve of w continues to not approve of w. While the preceding explains why either of these constitutes a monotonicity violation, neither situation fits our intuitive sense of a candidate losing when they gain votes. A similar situation leads to a violation of the independence of irrelevant alternatives: again, while such a violation could occur, it would require a voter raising the rank of a third candidate while *not* approving of a formerly winning candidate.

Strictly speaking, a pure approval system would allow voters to choose as many or as few candidates as they wanted. But how many candidates do voters choose? In 1985, TIMS (The Institute of Management Science), a professional society with about 6,000 members, performed an experiment. The annual election of its officers was by a traditional plurality vote, where the candidate with the most votes won. TIMS also asked its members to submit a nonbinding approval ballot, indicating which of the candidates the member approved of. The TIMS data suggest that, given the choice, voters tend to approve of only two or three candidates.[6]

The TIMS elections had at most five candidates, so approval of just two or three might reflect the limited number of choices and not limits on how many candidates a voter is willing to approve of. A more significant experiment occurred in 2002 during the elections for the French president. About 2,500 French voters were asked to submit non-binding approval ballots. This election had 16 candidates, and voters could approve of any number of them, but on average, voters chose slightly more than three candidates.

6. Based on its results, TIMS adopted approval voting for the election of its officers. In 1995, TIMS joined with the Operations Research Society of America (ORSA) to become INFORMS (Institute for Operations Research and Management Science), a society with about 12,000 members; approval voting has been retained for the new organization.

This suggests the following approach: suppose that voters were presented with a list of candidates, and could pick up to three of them to approve of; the candidate who received the most votes would be declared the winner of the election. The empirical evidence suggests that the votes submitted in this "pick three" system would not differ too greatly from the votes submitted in a pure approval system.

Approval voting is used by a number of professional societies, including INFORMS, and the MAA (Mathematical Association of America), the AMS (American Mathematical Society), the ASA (American Statistical Association), and IEEE (Institute of Electronics and Electrical Engineering); in addition, many universities and colleges now use approval voting for electing university officers.

One problem with widespread adoption of approval voting in the United States is historical: a form of approval voting *had* been used, and then abandoned. In its original form, the Electoral College chose the U.S. president and vice president by approval vote. Article II, Section 1 states: "The Electors shall meet in their respective States, and vote by Ballot for two Persons, of whom one at least shall not be an Inhabitant of the same State with themselves . . . The Person having the greatest Number of Votes shall be the President, if such Number be a Majority of the whole Number of Electors appointed." Thus, except for the restriction that an elector must cast at least one vote for an out-of-state candidate, the original form of the Electoral College was a "pick two" approval vote system. Unfortunately, the election of 1800 led to a tie between Thomas Jefferson and Aaron Burr, both Republicans, and the House of Representatives decided the victor—after 35 indecisive ballots.

The impasse in the House was caused by politics, not by approval voting. Nevertheless, Congress passed and the states approved the 12th Amendment, which established the Electoral College in its current form by abandoning approval voting. The 12th Amendment is an explicit rejection of approval voting in at least one context, and could be used to prevent approval voting in other contexts.[7]

---

7. Since the 12th Amendment still refers ties to the House of Representatives, it does not prevent a recurrence of the events of 1800. Congress's reaction is like the patient who, upon receiving bad news from the doctor, changes physicians.

Approval Voting and Multimember Districts

The "pick three" system described earlier bears some functional similarity to what is known as *plurality at large* voting. This is used in some states to elect representatives to the legislature using what are known as *multimember districts*. For example, representatives in Arizona's lower house are elected in two-member districts: voters in each state district vote for two candidates for office, and the two highest vote-getters are elected. In 1984, Gary Cox of the University of Texas at Austin pointed out that such a system is essentially an approval vote.

As before, imagine the voters and candidates distributed in a one-dimensional policy space. If voters pick two candidates and vote sincerely, they will select the two candidates closest to their position in the policy space. For example, if the voters and candidates are distributed as in figure 1, then voters A, B, C, and D would pick w and x; voters E and F would pick x and y; and voters G through L would pick y and z. Thus w would receive 5 votes; x would receive 6; y would receive 7; and z would receive 5 votes. As a result, x and y would win the election.

Note that the two candidates represent different ends of the political spectrum. The theoretical underpinnings of this result are the following. Imagine that the voters in a two-member district are distributed along a one-dimensional policy space. Suppose each party can run two candidates. Where should the candidates be placed so that the party is best positioned to win both seats?

It should be clear that at least one candidate should be placed as near as possible to the center of the distribution. Suppose one party places a candidate in the center. The second party cannot afford to cede the center, so one of their candidates should also be placed close to the center. This leads to a result known as the *median voter theorem*: successful candidates are those nearest to the center of voter distribution in the policy space.

Where should the parties place their second candidate? It's tempting to reason as follows: If a candidate wins when he or she sits in a certain spot in the policy space, then a second candidate, placed in the same spot, will also win, because everyone who voted for the first candidate will also vote for the second candidate. But this would be wrong!

The problem is the votes for a candidate in a "pick two" system come from those who ranked that candidate first, as well as those who ranked that candidate second. For example, suppose x's party ran a second candidate x', whose position in the policy space was the same as that of x. Under the original placement of the candidates, x won 6 votes and y won 7.

Under the new placement of the candidates, voters B, C, D, and E would pick x and x'. Voter A would pick w and *one* of x or x'. Likewise, voter F would pick y and *one* of x or x'. Thus the two votes that originally went to x are now split between x and x'. This means that, *at best*, x wins 6 votes (in which case x' wins 4 votes). But if A picks x and F picks x' as their second choice, then the two candidates win 5 votes apiece.

In contrast, y wins six votes: the first-place votes of F, G, and H, and the second-place votes of J, K, and L. z wins five votes: the first-place votes of J, K, L, and the second-place votes of G and H. The real loser is w, who would win the first-place vote of A, and *no other votes*: voters like B and C, who would have given their second-place votes to w, now give them to x' instead.

The best possible outcome, if the party of x ran a second candidate x' in the same part of the policy space, is that x or x' wins. But if the two candidates split the vote, each winning five apiece, then y still wins the first seat, and there is a three-way race tie for the second seat between x, x', and z. If we ran a run-off election where each voter chose one candidate, z would win the race, with six votes; meanwhile, x and x' would split five votes between them. As a result, the winners would be y and z, which would shut out x's party completely!

It follows that x's party should place its second candidate in another part of the policy space. This suggests that the candidates successfully elected from multimember districts will be more diverse than candidates elected from single-member districts: this is known as the *extremism hypothesis*. A 2008 study of the Arizona legislature, by Anthony Bertelli of the University of Georgia and Lilliard E. Richardson, Jr., of the University of Manchester in the United Kingdom, confirms this prediction: the candidates who are elected tend to represent a broader range of viewpoints than those elected from single-member districts. Arizona was a particularly useful study in this sense, since a comparison could be made directly: the same districts were simultaneously multimember districts, electing two state representatives, and single-member districts, electing one state senator.

Because of this, a useful comparison could be made. Consider the Arizona state senator. By the median voter theorem, this senator represents the median voter in the district. But the same voters elect two state representatives. It follows that the position of these representatives in the policy space can be compared to the position of the senator from the same district; and the difference between their positions and the senator's position can be used to measure of the validity of the extremism hypothesis.

How does this apply to federal elections? Since the Apportionment Act of 1842, all congressional districts have had voters elect one (and only one) representative: they are single-member districts. But in 1932, Minnesota became a laboratory for democracy.

Minnesota had long been dominated by Republicans; in fact, no Democrat had won a statewide or congressional election since 1918. However, changing demographics and the onset of the Great Depression produced a significant partisan shift and the rise of a new party, the Farmer-Labor, which threatened Republican hegemony. In 1930, the Farmer-Labor candidate Floyd Olson soundly defeated Republican rival Ray Chase, 57.1% to 35.0%. The Democratic candidate, Edward Indrehus, ran but did not campaign, taking 3.5% of the vote.

If each congressional district had a similar distribution of voters, Farmer-Laborites could win in every one of Minnesota's 10 congressional districts. They won one, but in three other districts, the Farmer-Labor candidates took more than 46% of the vote. Fortunately for the Republicans, they still controlled the legislature, and the 1930 census reduced Minnesota's congressional delegation from 10 to 9, forcing the state to redistrict.

The Republican-dominated legislature submitted a blatantly partisan plan. Many members of the Farmer-Labor Party found themselves "packed" into as few districts as possible, one of which had 344,500 people. Meanwhile the city of Minneapolis, a Farmer-Labor stronghold, was "cracked" into several districts, one of which reached 175 miles west to the Minnesota-Dakota borders; this dilution of the Farmer-Labor voters would make it harder for them to win the district. Republican incumbents, on the other hand, found themselves in smaller districts dominated by Republican voters: one such district had 228,596 residents.

Olson vetoed the plan, leading to a legal fight that wound its way to the Supreme Court. In *Smiley v. Holmes* (1932), the Court ruled that the governor's veto was legitimate; and that failure to implement a redistricting plan required the state to elect its representatives "at large." In effect, the entire state of Minnesota became a single 9-member district.[8]

---

8. This does not violate the requirement of single-member districts, since the Apportionment Act of 1842 provides exceptions for when a state fails to redistrict.

Here the threat to the Republicans loomed even greater than before. In the 1932 election, every Minnesota voter would vote for up to nine candidates; the top nine vote-getters would be elected. The Farmer-Laborites, who elected Olson by a sizable majority, would be able to win all nine congressional seats if they voted as a bloc: "Now for a clean sweep!" became the rallying cry for the progressives.

The possibility of a sweep was even greater than it might appear, since the Farmer-Laborites need not even have a majority: it is enough for them to have a plurality. If Farmer-Laborites made up 38% of the electorate, Republicans 36%, and Democrats 25%, then if voters only voted for candidates from their party, every Farmer-Labor candidate would win 38% of the vote, every Republican candidate 36%, and every Democratic candidate 25%. As long as there were at least nine Farmer-Labor candidates (and there were), they would win all of the available seats.

This was exactly the distribution of votes in the 1932 elections. But voters did not vote only for candidates from their own party (hardly a surprise, since the same voters that elected a Farmer-Labor governor also elected Republican state legislators and representatives). The Farmer-Labor party did not win all nine seats—they won five, the Republicans three, and the Democrats one.

By any reasonable standard, the Republicans got their fair share of the congressional seats: with about one-third of the vote, they won one-third of the seats. The Farmer-Labor Party did gather more than their fair share, by taking more than half the seats with just 38% of the vote. However, the Democrats and Farmer-Laborites formed a de facto progressive coalition that won 67% of the seats with 63% of the votes. By any reasonable standard, the 1932 Minnesota congressional delegation produced the fairest possible outcome given the distribution of popular votes.

The Minnesota experiment only lasted one election cycle. Shortly after, the legislature submitted a viable redistricting plan, and the governor approved it. In the 1934 election, Minnesota was divided into nine single-member districts. This time, Republican candidates won one-third of the vote—but more than half the seats.

Dragons and Dummymanders

> The Times, Places and Manner of holding Elections for Senators and Representatives, shall be prescribed in each State by the Legislature thereof; but the Congress may at any time by Law make or alter such Regulations, except as to the Places of chusing [*sic*] Senators.
>
> —ARTICLE I, SECTION 4

The 2000 census reduced Pennsylvania's congressional delegation from 21 to 19. As a result, the state had to redraw the district lines. Because state legislatures decide on districting plans, the party that controls the legislature at the time of the census (the "in-party") has an inordinate amount of power: it can draw the district lines to favor itself to the detriment of its opponents (the "out-party" or -parties). This produces a *partisan gerrymander*, named after Massachusetts governor Elbridge Gerry, who supported an 1812 districting plan with a bizarrely shaped district designed to give his party (the Democratic-Republicans) a decided advantage in state legislative elections.

In 2000, Republicans controlled both houses of the Pennsylvania legislature. They also controlled the governorship (thereby preventing a repeat of the Minnesota non-redistricting) and were doubtless emboldened by the fact that they also controlled the presidency and both houses of Congress. To divide their state into congressional districts, Pennsylvania Republicans enlisted the help of national Republican strategist Karl Rove; Speaker of the House Dennis Hastert of Nebraska; and House Majority Whip Tom DeLay of Texas.

Traditional partisan gerrymandering involves a strategy known as "packing and cracking." First, pack as many of the out-party supporters into as few districts as possible. Next, crack concentrations of out-party supporters so that they are split between several districts, making it harder for them to win any districts. For example, the original gerrymander came from cracking a Federalist district.[1] With the new redistricting plan, Republicans won 12 seats (one more than they had in 2000) and Democrats, 7 (three less).

Almost immediately, Democrats filed suit. But in *Vieth v. Jubelirer* (2004), the U.S. Supreme Court ruled against them, citing the lack of a "manageable standard" for measuring the extent of partisan gerrymandering. This gave a green light to blatant partisan gerrymandering, and a sufficiently brazen in-party could effectively eliminate the out-party through legislative fiat.

The Whole of the Law

The Court's decision implies that if a manageable standard for measuring the extent of partisan gerrymandering could be found, it might be willing to revisit the question. Here, we consider measurable features of congressional districts. Since the number of congressmen is based on a state's population, we might begin by requiring that all congressional districts have the same number of people. But as we saw in chapter 1.22, this is impossible, for congressmen in different states will in general represent different numbers of people.

At the very least, we might require that the congressional districts *within* a state have the same number of people. As remarkable as it might seem, the idea that congressional districts within a state should represent equal numbers of people did not become a requirement until the 1960s! In Tennessee, state legislative districts had been drawn up in 1901 and not changed for 60 years, though in the meantime the state's demographic characteristics had been radically altered by two world wars and the Great Depression. In *Baker v. Carr* (1962), the Supreme Court ruled that the failure of the state of Tennessee to redraw state legislative districts in a timely manner, which led to widespread population inequities, was *justiciable*, which means that the courts could intervene. *Reynolds v. Sims* (1964) went further, establishing that, as nearly as possible, state legislative districts should contain equal numbers of people. Chief Justice Earl Warren went so far as to claim *Reynolds v. Sims* as

---

1. Ironically, it failed: the Federalist party held the seat, until its opposition to the War of 1812 caused the party to lose favor with the electorate.

his most important opinion—even more important than *Brown v. Board of Education* (1954)!

The Court first applied the equipopulation principle to congressional districts in *Wesberry v. Sanders* (1964), where Georgia's 5th congressional district had a population of more than 800,000, twice the average size of Georgia's other districts (and nearly three times the size of the smallest of Georgia's congressional districts).

Wesberry v. Sanders, at least, established what seemed to be an ironclad measure of districting fairness: equal population. But even this slight hope of challenging a districting because of population inequalities vanished with *Karcher v. Daggett* (1983), which addressed redistricting in New Jersey. There the Supreme Court held that "consistently applied legislative policies might justify some variance." These policies might include respecting municipal boundaries, maintaining previously existing districts—and protecting incumbents. Since the in-party necessarily has a number of incumbents, *Karcher v. Daggett* gave the go-ahead to provide them with so-called safe districts: districts whose populations leaned so heavily in favor of the in-party that the in-party candidate was nearly certain to win. The whole of the law is that congressional districts should have equal numbers of people—except when they don't.

Measuring Shape

The fact that equipopulation is not enough is clear from Pennsylvania's bizarrely shaped districts, which, like the clouds in a summer sky, can be used as an exercise in imagination: Justice Stevens described Pennsylvania's 6th district as a "dragon descending on Philadelphia from the west." This observation implies that congressional districts should *not* look like dragons; indeed, the eponymous bizarre district was caricatured as a salamander. But if districts should not look like amphibians or mythical beasts, what *should* they look like? Of all the states in the country, Iowa alone identifies the ideal shapes for districts: "In general, reasonably compact districts are those which are square, rectangular, or hexagonal in shape."[2] However, there is a problem: the

2. Iowa State Law, Code Section 42.4(4).

equipopulation requirement, combined with the fact that most state boundaries are irregular, generally will make it impossible to divide a state into squares, rectangles, and hexagons. Thus, some deviation from these figures must be expected. The important question is how that deviation will be measured.

Iowa's statute refers to the idea of *compact* districts; in fact, many observers identify compactness as a key component of a properly constituted district. But what is compactness? Mathematicians and political scientists have presented more than a dozen ways to measure this property, and some of these measurements have worked their way into state law. For example, Iowa uses the ratio between the length and width of a district as one measure of compactness. Unfortunately, this measure only works when districts have a clearly defined length and width; trying to apply this to Arizona's 2nd congressional district (see figure 3) would be challenging![3]

We must have some other method of measuring how far a district's shape is from the ideal. Because it may be possible to obey the letter of the law while breaking its spirit, we should consider our measures of compactness carefully. In particular, if we hope to use a measure of compactness to evaluate the extent of partisan gerrymandering, then we must measure an unavoidable consequence of partisan gerrymandering. To that end, we must figure out how we can produce a partisan gerrymander.

Since it will generally be impossible to divide a state into congressional districts of equal size or shape, let's consider what *one* congressional district should look like. The most regular figure possible is a circle, so we may consider the ideal district to be circular, and base our measure of compactness on how a district deviates from circularity. Here's the first problem: which circle should we compare a district to? Since we want to measure something that will occur when a partisan gerrymander is produced, we might approach the problem from the opposite direction: Suppose we begin with a perfectly circular congressional district. How would a partisan gerrymander affect the district's shape?

There are two ways to proceed. Although there are two strategies for partisan gerrymandering, we note that packing the out-party into just a few districts will generally crack them in the remainder. There are also theoretical

---

3. The 2nd district has been redrawn for the 113th Congress and has a slightly less bizarre appearance, though it still loops around a substantial portion of the state.

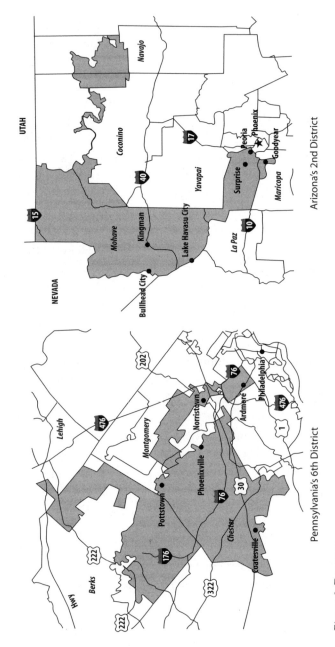

Arizona's 2nd District

Pennsylvania's 6th District

Figure 3. Dragons

reasons that suggest cracking is not in general a good strategy, so we will focus on the packing strategies available to the in-party.

Suppose we began with a perfectly circular district that encloses the right number of people to satisfy the equipopulation requirement. If we wanted to pack this district with members of the out-party, we cannot simply expand the district to include them, as that would violate the equipopulation requirement. Instead, we must trade members of the in-party in the district for members of the out-party outside of the district: this trade is termed a *voter exchange*.

To do this, consider a member of the in-party who lives just inside the district, and a member of the out-party who lives just outside the district. A gerrymanderer would dimple the district line inward, so the in-party resident now lives outside the district. At the same time, the gerrymanderer would push the district line outward (pimple?) to include the member of the out-party. In this way, the population of the district remains the same, but the number of members of the out-party increases.

To a first approximation, the dimples and pimples have the same area, so the total area of the district does not change. But its perimeter does change: dimples and pimples increase the perimeter of the district.[4] We might measure the district's shape by comparing its perimeter to the circumference of a circle with the same area (the *isoareal circle*). We can make the comparison by forming the ratio between the perimeter of the district and the circumference of the circle with the same area. This is the basis of the *Schwartzberg measure*, whose use was suggested in *Wiser v. Hughes* (1982). (The lawsuit was dismissed as involving no federally justiciable question.)

We might go further. However bizarrely shaped a state's congressional districts might be, it seems petulant to attack a districting plan without providing an alternative plan that is demonstrably superior. Since every voter must be in a congressional district, the dimple of one district must accompany a pimple of another—and so the perimeters of both districts will increase. Thus we might compare two districting plans by considering the sum of the perimeters of all the districts. Colorado has incorporated this idea

---

4. This is not, strictly speaking, true: a sufficiently small dimple would, by itself, decrease the perimeter. However, if we regard dimpling and pimpling as occurring simultaneously, and if we hold that the dimples and pimples are of equal area, the increase in perimeter caused by the pimple would more than offset the decrease in perimeter produced by the dimple.

into its constitution, while Iowa uses it as a second way to measure district compactness.

As we might expect, finding the perimeter is challenging, especially if the district twists and turns, leading us to try a different approach to measuring the magnitude of a partisan gerrymander. Again, suppose we want to produce a packed district. This time, we begin with a perfectly circular region that contains enough out-party voters to meet the population requirement. The circle has the ideal shape for a district, but it will very likely contain more voters than a district should. Thus we would produce a packed district by removing the in-party voters from the district. We can use the fraction of the area of the circle remaining as a measure of how gerrymandered the district is. Equivalently, we take the circumscribing circle (the smallest circle that completely encloses the district) and find the fraction of the area enclosed by the circle that is part of the district. This is the basis for the *Reock measure.* Michigan uses a modified Reock measure: first, only the land area in the circle is used; and second, it measures the fraction of the land area of the circle excluded from the district.

The *Angel-Parent measure* combines both the voter-exchange concept of the Schwartzberg measure with the fraction-of-area concept of the Reock measure. First, we take an isoareal circle, centered at the district's geographic center. As with the Schwartzberg measure, we may take this as the circle that contains enough population to form a congressional district. Now we exchange voters inside the circle for voters outside the circle. For example, if we want to pack the out-party, we trade concentrations of in-party members inside the circle for concentrations of out-party members outside the circle. The area within the isoareal circle but not within the district corresponds to the voters who have been exchanged for voters outside the circle; we use this fraction as the Angel-Parent measure.

There is another possibility—that we draw a district with the requisite numbers of out-party and in-party voters. If we imagine the district boundary to be a fence that we have erected around a group of voters, we might ask if we could have drawn a *different* district using a fence of the same length. Under the assumption that the ideal district should be circular, we might compare the district's area to the area of a circle whose circumference is equal to the district's perimeter: the *isoperimetric circle.* This is the basis of the *Cox measure.*

We could abandon the notion of an ideal shape altogether. Imagine we draw the district's boundary by beginning at some point and moving in a straight line to a nearby point; suppose we are moving in such a way that the district is always to the left of our line. Because the boundary line must eventually return to where we start, we must necessarily turn to the left ("toward" the district) a number of times. However, if we are packing members of the out-party, we will occasionally turn to the right ("away" from the district) in order to include some members of the out-party. This forms a *reflexive angle*. The other angles, formed when we turn toward the district, are *nonreflexive angles*. The *Taylor measure* is the difference between the two, divided by the total number of angles.

One problem with the Taylor measure is that it assumes district boundaries are straight lines; a curved district boundary would not produce any angles, reflexive or otherwise. We might instead consider the *convex hull* of the district: this is the shape of least area with no reflexive angles that contains the district. This is sometimes called the "rubber band" shape, since we can form it by stretching a rubber band around the district boundary: the rubber band would run along the boundary as long as the boundary turned toward the district, but every time the boundary turned away, the rubber band would be pulled away from the boundary line. As with the Reock measure, we might consider the area of the district as a fraction of the area of the convex hull.

We raise three objections to the preceding measures. First, the irregularities of politics and geography guarantee that these ideal district circles will often contain regions where no state voters live: bodies of water and portions of other states or countries. The Angel-Parent, Reock, and convex hull measures can be easily modified to account for this by excluding these areas from the computation; indeed, Michigan's variation on the Reock measure does exactly this.

A more serious objection is that congressional districts are defined by populations, not geography. Again, it is easy enough to modify the Angel-Parent, Reock, and convex hull measures to account for this, since we can use the eligible voters inside the ideal figure instead of the area of the figure (and for the Angel-Parent measure, we can use the population center of the district).

For example, if we enclose a district with a circumscribing circle, the standard Reock measure would be the fraction of the area of the circle represented by the district. A modified Reock measure would be the fraction of the *voters* in the circle that are members of the district. A similar modification could be made for the Angel-Parent and convex hull measures.

The Schwartzberg and Cox measures are somewhat more difficult to modify: they do not specify the location of the ideal circle, so the number of eligible voters inside this circle will vary. The Taylor measure would be unaffected by whether we measure populations or areas.

The Best Measure of Gerrymandering

Which of these measures is best? The answer ultimately depends on what we want to do with the measure of gerrymandering. Legally, there are two applications. First, we might introduce a shape measure in order to argue that a district has been improperly drawn. While *Vieth v. Jubelirer* established that partisan gerrymandering was permissible, *racial* gerrymandering is not.

In 1965, Congress passed the Voting Rights Act. To ensure that persons were not denied the right to vote, the act required certain jurisdictions to obtain approval from the federal government prior to making any changes in the election process. Fast forward to the present: in one of the most controversial decisions of 2013, the Supreme Court struck down the preclearance requirement, on the basis that the listed jurisdictions had changed since 1965. This is tantamount to claiming that voter discrimination is no longer practiced in the United States.

The 1990 census increased North Carolina's congressional delegation by 1 seat, to 12, so the state drew up a new districting plan. Since 40 of North Carolina's 100 counties were jurisdictions covered by the provisions of the Voting Rights Act, the state submitted its plans to the Justice Department for preclearance.

At the time, North Carolina's population was 78% white and 20% black. Moreover, the black population was dispersed: only 5 of the state's 100 counties had a majority black population. The simplest possible geographic dissection of the state into congressional districts would result in blacks being a minority in all congressional districts. To prevent this, the state created a *majority-minority district*, where the majority of the population belonged to a minority group.

However, the Justice Department ruled that the North Carolina plan did not go far enough: since blacks constituted 20% of the population, they should be able to win a like proportion of the districts in the state. The U.S. attorney general's office noted that it appeared feasible to create a second majority-minority district in the south central to southeastern part of the state; in fact, one of the plans rejected by the North Carolina legislature included such a district.

The state responded by creating a second majority-minority district, the 12th, but not in the area suggested by the attorney general. Rather, the 12th district wound 160 miles through the central part of the state, mostly along I-85; one commentator quipped that a person who drove along I-85 with both car doors open would kill most of the residents of the district. This is an exaggeration: sometimes only the northbound lanes were in the 12th district, and sometimes only the southbound lanes were in the district. The other majority-minority district, the 1st, was likened to a "bug splattered on a windshield."

It appears that the legislature's rationale for creating such a bizarrely shaped district were twofold. First was the Justice Department's implicit requirement that a second majority-minority district be created. This task was complicated by the dispersal of North Carolina's black population, so constructing *any* district to ensure that it had a majority black population would require some wild distortion of sensible geography. Second was a desire on the part of the legislature to protect current incumbents, which was explicitly recognized in *Karcher v. Daggett* as a legitimate reason to distort district lines.

The two majority-minority districts, the 1st and the 12th North Carolina congressional districts, went into effect for the 1992 congressional elections, and it was probably not coincidence that, for the first time since the end of Reconstruction, North Carolina elected black representatives: Mel Watt in the 12th district and Eva Clayton in the 1st.

However, a group of white voters in the 12th district launched a suit, charging that the district had been drawn *because* of the racial makeup of its residents: it was a racial gerrymander and thereby prohibited under the equal protection clause of the 14th Amendment.

The Supreme Court, in *Shaw v. Reno* (1993), issued a carefully worded decision. First, it meticulously avoided the question of the constitutionality of racial gerrymanders. It also avoided the question of the constitutionality of gerrymanders. What it rejected was districts so bizarrely shaped that the only reasonable explanation for their shape was racial gerrymandering.[5]

---

5. In response, North Carolina redrew the 12th district, making it harder for an errant interstate driver to wipe out the district's population. Mel Watt would be re-elected as the Congressman for the 12th district for the eleventh time in 2012.

The key word here is "bizarre." Most gerrymandering cases express a desire for *compact* districts, and in general, noncompact is equated with bizarre. Unfortunately, the Court has defined neither bizarre nor compact, but relies on the "I know it when I see it" standard. We might wonder why the Court has consistently refused to identify a preferred method of measuring compactness. Part of the reason is that any measure of compactness invariably produces examples that are quantitatively equivalent, but offend our intuitive sense of compactness.

For example, consider the Reock measure. An equilateral triangle inscribed in a circle takes up about 41% of the area of a circle and meets most people's intuitive sense of what a compact district should look like. But another shape, that also takes up 41% of the area of a circle, would also be considered compact by the Reock measure, no matter how bizarre its boundaries. This is the virtue of having multiple measures of shape. In general, consider *any* shape someone would regard as noncompact. It will fail at least one of the shape measures; our task is to identify which one it fails.

We might be uncomfortable with the idea of *finding* a measure that a noncompact district fails; this seems too much like changing job requirements to exclude a specific candidate. But *all* candidates for a job must meet *all* requirements; the failure to meet a single requirement is sufficient to reject a candidate. In the same way, a district that fails even a single compactness test should be regarded as noncompact.

The Art of the Dummymander: A Self-Limiting Enterprise?

In *Bandemer v. Davis* (1986), the Supreme Court ruled that partisan gerrymandering *was* justiciable. Justice O'Connor suggested that, even so, court intervention might not be necessary: a partisan gerrymander necessarily requires weakening the in-party's advantage in safe districts.

Empirical evidence for Justice O'Connor's viewpoint can be found in events following the 1990 census. In several southern states, Democrats controlled the redistricting process. Since every district packed is a district effectively abandoned to the other party, the Democrats packed as few Republican districts as possible, giving themselves narrow margins in other districts: 51% to 49%, for example. But such a narrow margin can easily be eroded by partisan shifts, and in the 1990s, the south began turning Republican. As a result, a district that would have returned a narrow Democratic victory in 1990 became a district that returned a narrow Republican victory in 1992.

The partisan gerrymanders backfired so catastrophically that they introduced a new term to our political vocabulary: the dummymander. It's not clear to what extent the southern dummymanders influenced the *Vieth* decision. But the fact that unrestricted partisan gerrymandering carried its own punishment would hardly have escaped the justices in *Vieth*. However, Justice O'Connor's view misses a crucial point: partisan gerrymandering can be self-limiting—but what is the limit? Mathematicians studying *game theory* answer this type of question as follows.

One of the basic techniques of game theory is finding the *expected value* of a choice (*strategy*): this is the benefit (*payoff*) that could arise from this choice, multiplied by the probability of receiving this payoff, minus the cost of making that choice. For example, consider a lottery, where $1 buys you a ticket that has one chance in ten of winning $5. There are two strategies: buy a ticket, or don't buy a ticket. If we buy a ticket, we could win the payoff of $5 with probability 1/10. However, the strategy of playing costs us $1. Thus the expected value of this strategy is $5 \times 1/10 - 1 = -0.50$. On the other hand, if we don't buy a ticket, we have a guaranteed payoff of $0, and a strategy cost of $0, for an expected value is 0. Since the expected value of the strategy "buy a ticket" is less than the expected value of the strategy "don't buy a ticket," the rational decision is to not buy a ticket.

We might wonder why people buy lottery tickets, since every lottery in existence has a negative expected value. The simplistic answer is that people don't make rational decisions. But the correct answer is that people *do* make rational decisions—the apparent irrationality of buying a ticket results from limiting our notion of payoff to the actual cash payout.

Those who run games of chance understand this intuitively, and offer prizes worth more to the winners than their cost to the house. Money from the lottery goes to education; or you qualify for a player's club; or bells and whistles go off in a grand fête when you win; or you get free drinks while you play. In some cases (money for education or free drinks), you win these benefits with certainty; in other cases (bells and whistles), these benefits increase the payoff when you win. While we cannot assign an objective monetary value to these things, the players themselves assign subjective values to them and we can, with properly designed experiments, find these values. In cases like a prize of free drinks or education money, the subjective value to the player is often far more than the objective cost to the house: in other words, what the players thinks these perks are worth is generally far more than what the house or state actually pays for them. Such payoffs exist even for pathological gamblers: neurological research indicates that addicts receive a "high" while engaging in their addiction, which increases the payoff.

Consider gerrymandering. The strategies available to the in-party are the possible districting plans. We can then compute the expected value for each plan. The *optimal strategy* is the one that has the highest expected value.

For example, suppose a state is to be divided into 10 districts, and that the two parties have equal presence among the electorate; in other words, each party constitutes 50% of the registered voters. By any reasonable standard, a fair division would give each party five of the ten available seats. If game theory predicts that the optimal strategy leads to this division, then gerrymandering is a self-*regulating* enterprise, and the Supreme Court might as well declare it nonjusticiable: too blatant a partisan grab would be its own punishment. On the other hand, if game theory predicts that the optimal strategy leads to a division that is far from fair, then legal intervention is desirable.

Let the gerrymandering strategy be the following: First, draw up one or more districts that contain only members of the out-party: these are the packed districts. Then divide the remaining members of the out-party among the remaining districts: these are the cracked districts. The out-party wins the packed districts by landslides and loses the cracked districts by narrow margins. Since there are only enough members of the out-party to pack five districts, we can determine the number of seats the in-party can expect to win when the out-party is packed in 1, 2, 3, 4, or 5 districts, then use the strategy that gives the in-party the most seats.

For example, the in-party could pack one district completely full of out-party voters, then crack the remaining out-party voters by dividing them evenly among the remaining districts. In the first district, the out-party would constitute 100% of the electorate, while in the remaining districts, they would make up 44% of the electorate. If we assume absolute voter turnout and absolute voter loyalty, the strategy "pack one district" leads to an expectation of 9 districts won by the in-party. It's easy to determine that the "pack two districts" leads to an expectation of 8 districts; that "pack three districts" leads to an expectation of 7 districts; and so on. Thus the "pack one district" is the optimal strategy for the in-party.

Of course, it would be exceedingly naïve for the in-party to assume absolute voter turnout and loyalty (though evidently this is what occurred in the southern dummymanders!). It is more sensible to assume that only a fraction of in-party voters are loyal. Suppose that both parties have equal presence in the population, in the sense that equal numbers of voters profess to support the two parties. A little algebra shows us that if L% of the districts are cracked, then party loyalty must be more than L% in order for in-party candidates to win in the cracked districts.

Suppose we knew $L = 80\%$: In other words, we knew that exactly 80% of in-party voters would vote for in-party candidates. Then we could crack up to 80% of the districts and win all of them. Since party loyalty was 80%, and the in-party made up 50% of the electorate, it follows that the in-party could win as much as 80% of the districts with only 40% of the vote. This should be disturbing. However, the situation might be saved by the fact that the strategy of cracking nearly 80% of the districts would risk perfect dummymander, since if party loyalty were even a little less than 80%, the in-party would lose *all* the cracked districts and, as they have no chance of winning the packed districts, they will win 0% of the seats! This is the mathematical basis for a belief, like that of Justice O'Connor's, that partisan gerrymandering would be self-limiting.

Since the actual party loyalty is not known in advance, we can treat it as a random variable with some mean and standard deviation. Suppose we use poll data to determine that in-party voters have a mean loyalty of 80% with a standard deviation of 5%. *Chebychev's theorem* tells us that the probability that party loyalty is within k standard deviations of the mean is at least $1 - 1/k^2$. For example, the probability of having a mean loyalty within $2 \times 5\% = 10\%$ of the mean, that is to say between 70% and 90%, is at least $1 - 1/2^2 = 75\%$. Put another way: There is *at least* a 75% chance that in-party voter loyalty will be at least 70%.

Now suppose we crack 70% of the districts. We need a party loyalty of more than 70% to win a cracked district. Chebychev's theorem guarantees that this will happen at least 75% of the time.[6] Thus the in-party can expect

---

6. Strictly speaking, if party loyalty is exactly 70%, the in-party ties in the cracked districts. We could eliminate this chance at the expense of making a somewhat more complicated statement: thus, Chebychev's would guarantee that there is at least a 74.95% chance that party loyalty will be at least 70.01%.

to win 75% of the cracked districts. Since the in-party has no chance of winning the packed districts, it can reasonably expect to win at least 52.5% of the total districts: that is, 75% of the 70% of the districts that are cracked.

What if we cracked 65% of the districts? In this case, party loyalty must exceed 65%, and Chebychev's guarantees that this will happen at least 89% of the time. The in-party can expect to win at least 57.8% of the total districts: again, 89% of the 65% of the districts that are cracked. In a similar fashion, we can find our expectation if we cracked 60% of the districts, or 55%, or any other number. The *optimal strategy* will be the one that gives us the highest expectation. In this case, the optimal strategy involves cracking 65% of the districts, in which case the in-party can expect to win at least 57.8% of the districts.

In the sense that the in-party can win at best about 57.8% of the legislative seats, Justice O'Connor's intuition that partisan gerrymandering is self-limiting holds true. But remember that this was on the basis of an in-party with a mean loyalty of 80%, so that the in-party wins 57.8% of the districts with only about 40% of the votes.

In fact, there is a more ominous possibility. Suppose the in-party adopted platforms that appealed to a radical faction in the party, but that others found abhorrent. Party loyalty would drop, as those who found the new platforms appalling began to reconsider their support for the in-party; at the same time, those who remained would be even more devoted to the party. This corresponds to a decrease in the mean loyalty and the standard deviation. Thus the party might evolve from one with a mean loyalty of 80% and a standard deviation of 5%, to one with a lower mean loyalty and lower standard deviation: say 65% mean loyalty with a 1% standard deviation.

Under these conditions, the optimal strategy has the in-party cracking about 60% of the districts, which gives it an expectation of winning about 57.6% of them. While this is almost the same percentage as before, the lower mean loyalty of 65% means that the in-party wins 57.6% of the districts with just 32.5% of the vote.

LULAC v. Perry

As bad as the preceding situation appears, the reality is even worse. Texas underwent a redistricting after the 2000 census. But when Republicans took control of the legislature in 2003, they instituted a *second* mid-decade redistricting. In the post-*Vieth* world, the plan's architects did not even attempt to conceal the partisan purpose of the redistricting. Some concentrations of Latino voters were cracked, and the League of United Latin American Citizens (LULAC) sued Governor Rick Perry over the redistricting.

The decision in *Vieth* suggested that it was the lack of a manageable standard of measuring partisan gerrymandering that made it a nonjusticiable issue. To remedy this, a group that included Gary King of Harvard University and Bernard Grofman of the University of California at Irvine, presented an *amicus curiae* briefing to the Court, noting that political scientists had long been using *partisan symmetry* as a measure of the fairness (or unfairness) of a districting plan.

Suppose, as above, a political party can win 57.6% of the districts with just 32.5% of the vote. This by itself is not objectionable if it held true for *any* political party; it would only be objectionable if it held true for some political parties and not others. The difference between the percentages of the districts the parties would win with the same percentage of the vote can be used to measure the extent of partisan gerrymandering.

To that end, Ronald Gaddie, of Oklahoma State University, analyzed the Texas districting plan and concluded that if both parties had 50% of the vote, the Republicans would win 20 out of 32 (about 63%) of the congressional districts. In contrast, an alternate plan, also analyzed by Gaddie, would have given both parties about 50% of the districts with 50% the vote.

Partisan symmetry would seem to satisfy the Court's requirement of a manageable standard for measuring partisan gerrymandering. It lacks only a legal decision as to how much asymmetry could be allowed. Thus, even though the Texas plan would give the Republicans 63% of the congressional districts with 50% of the vote, the Court might have ruled that this partisan advantage is consistent with constitutional principles.

Instead, the Supreme Court explicitly avoided revisiting the justiciability question and issued an even more stunning decision: the Constitution did not limit how often redistricting could occur. In the nineteenth century, when redistricting required laying out maps and drawing district lines using markers, such a pronouncement might be reasonable: a state would be loathe to involve itself in the expensive and time-consuming process of redistricting more than it had to. But in the twenty-first century, sophisticated computer programs can produce a set of districts with any desired characteristics in seconds.

It is conceivable that the in-party could implement a redistricting immediately before *every* election. Under the preceding assumptions, it could still command a majority in the legislature long after it had lost the support of most of the electorate. And if it took advantage of the *Karcher* ruling, which allowed the protection of incumbents as a legitimate reason to violate the equipopulation requirement, it could maintain its grip even longer.

The Pathology of Democracy

Gerrymandering has been called the pathology of democracy: a bizarre construct that appears as a consequence of innocuous principles. In this case, there are three principles. First, every state is to be divided into a number of congressional districts, each of which elects a single person to the House of Representatives. Second, the election is by plurality, which means that the candidate who wins the greatest number of votes (not necessarily a majority) wins the district. And third, the candidates are supported by political parties.

We can eliminate gerrymandering by eliminating one of these. Eliminating the last would be problematic, since the 1st Amendment guarantees the right of free association. Even without this restriction, it would probably be impossible to eliminate: the support network of a candidate is ipso facto a political party.

Eliminating the second seems unlikely. While there are alternative electoral systems, like instant runoff or plurality-at-large voting, the prevalence of plurality voting combined with the perception (true or not) that it is the simplest system makes it unlikely that it will be replaced.

This leaves the first option: the single-member district. This was established by congressional fiat, and could be eliminated in the same way. The multi-member district, which is already used for some state legislative districts, is one possible alternative.

The concern with a multi-member district is that the majority party will simply sweep all available seats. This did not happen in Minnesota in 1932. However, other states that failed to redistrict and ran at-large congressional elections in 1932 did produce "clean sweeps." Thus the Democrats won all nine of Kentucky's congressional districts with 60% of the vote; Democrats won all thirteen of Missouri's congressional districts with 62% of the vote; Republicans won both of North Dakota's seats with 65% of the vote; Democrats won all nine of Virginia's congressional districts with 74% of the vote.

In 1962, Alabama Democrats won all eight seats with 80% of the vote; Democrats won both seats in Hawaii with 67% of the vote; both seats in New Mexico with 58% of the vote.

While this seems ominous, note that in every case the party that won the "clean sweep" secured a majority of the votes, and an unlimited partisan gerrymander would produce the same results. The fact that, in 1962, the Democrats won all eight of Alabama's congressional seats with 80% of the votes is unremarkable: if 80% of the vote in each of eight congressional districts went to the Democratic candidate, we would observe the same result.

Moreover, the study of Arizona politics, discussed in the previous section, seems to indicate that multi-member districts lead to greater *diversity* of political activity, regardless of party affiliation. This is because in a multi-member district, two candidates from the same party with identical viewpoints compete with *each other*, so it is in their best interests to have divergent (and, one hopes, honestly held) viewpoints. This competition with other members of the same party does not happen in single member districts, which thus encourages "party line" politics.

On the down side, empirical research shows that the candidates who are elected in multi-member districts tend to be less diverse *racially*. In particular, candidates from minority groups, particularly black and Latino, tend to do worse in multimember district elections than in single-member districts, though—intriguingly enough—women tend to do better.

Proportional representation is another alternative. Despite claims to the contrary, there is nothing unconstitutional about it; the only road block is the federal law requiring single-member districts. One way to implement it would be to have voters cast their ballots for a political party; then use the same method of apportioning congressmen to the states to apportion representatives to the parties. In principle there is no difference between apportioning 435 congressional seats among 50 states with differing populations and apportioning a state's congressional delegation among several parties with differing levels of support. There is one minor functional difference, in that every state is guaranteed at least one representative, while it would seem unreasonable to give a seat to political party that won only a few votes. Most systems of proportional representation require parties win a certain minimum percentage of the total vote before they are eligible for seats.

One objection is that proportional representation is like buying a car by choosing the make but allowing the dealer to choose the model. So, even if a voter supports a political party *in general*, he or she might object to a particular member of that party as their representative.

One way to circumvent this problem is to use *Single Transferable Vote*, which is used in a number of Australian states. As in the Borda count and instant runoff voting, voters rank the candidates. But once a candidate receives enough votes to be elected (the quota), additional votes for that candidate are transferred to the voter's second choice; and so on. If no candidate meets quota, the candidate with the fewest votes is eliminated and ballots including them are re-ranked; this process is repeated until a candidate reaches quota. The *Droop Formula* (named after an English lawyer) is used to determine the quota: It is $\dfrac{Total\ Votes}{Available\ Seats+1}$, rounded up to the next whole number (so if this is a whole number, it would be rounded up to the next); this guarantees that the number of candidates who meet quota will equal the number of seats available.

For example, if a 2-member district has 100 voters, the quota will be 34 votes (so by the time two candidates reach quota, there are only 32 votes remaining). If there are 4 candidates for 2 positions, we might receive the following ballots:

$a > b > c > d$, 20 voters
$a > d > b > c$, 20 voters
$b > c > d > a$, 17 voters
$d > a > b > c$, 18 voters
$c > a > b > d$, 25 voters

a, with 40 first-place votes, would be elected as one of the representatives. However, quota was 34, so the extra $40 - 34 = 6$ first-place votes for *a* should be transferred to second-place candidates. But which ones?

The most commonly used method is the following: Consider the second-place choices of the 40 voters who placed *a* at the top of their list. Half of the second-place choices were *b*, and half were *d*. Therefore, half of the excess votes should go to *b*, and half to *d*. The re-ranked ballots are:

$b > c > d$, $3 + 17 = 20$ voters
$d > b > c$, $3 + 18 = 21$ voters
$c > b > d$, 25 voters

Since no candidate meets quota, the candidate with the fewest first-place votes (b) is eliminated and the re-ranked ballots become:

$c > $ d, $20 + 25 = 45$ voters

$d > $ c, 21 voters

Thus c wins the second seat.

How does STV fare in comparison to other systems? Like most systems, it fails independence of irrelevant alternatives. It also fails monotonicity, in essentially the same way that instant runoff voting does: Raising a candidate's rank could cause a different candidate to be eliminated when none meet quota, changing the outcome of an election.

STV also fails the Condorcet criterion. Suppose we had 100 voters again, with a quota of 34, and ballots as follows:

$a > c > b > d$, 33 voters

$b > c > a > d$, 33 voters

$d > c > a > b$, 33 voters

$c > d > a > b$, 1 voter

c wins against a, b, and c (67 to 33 in each case). However, since no candidate meets quota, the candidate with the fewest first-place votes is eliminated—which means that c, the Condorcet winner, is eliminated immediately. Again, as with instant runoff, the Condorcet violation is immediately evident from the ballots submitted. Proponents of single transferable vote should take a lesson from the Burlington mayoral races, and legislation mandating its use should incorporate a clause giving victory to the Condorcet winners first, before the STV mechanism is invoked. ("Winners" because it's possible for there to be a Condorcet winner at *each* stage in the re-ranking process.)

The Worst Way to Elect a President, Except for All the Rest

Each State shall appoint, in such Manner as the Legislature thereof may direct, a Number of Electors, equal to the whole Number of Senators and Representatives to which the State may be entitled in the Congress.

—ARTICLE 2, SECTION 1

The Electors shall meet in their respective states and vote by ballot for President and Vice-President . . . The person having the greatest number of votes for President, shall be the President, if such number be a majority of the whole number of Electors appointed.

—AMENDMENT XII

In 1876, the Democratic presidential candidate Samuel Tilden won 50.9% of the popular vote, while his rival, Republican Rutherford Hayes, won only 47.9% of the vote. The outcome hinged on the electoral votes of three states: Florida, Louisiana, and South Carolina. But accusations of voter intimidation and confusing ballots raised questions about the validity of the vote in these states, and until this issue was decided, neither candidate had a majority of the electoral vote.

Congress appointed a 15-member commission consisting of 5 representatives; 5 senators; and 5 members of the U.S. Supreme Court. Originally there were to have been seven Democrats and seven Republicans, with the fifteenth member slot filled by David Davis, a Supreme Court Justice widely regarded by both parties as an Independent. However, at the last moment, the Illinois legislature appointed Davis to the Senate. Since five senators had already been assigned to the commission, Davis was replaced by Joseph P. Bradley, a Republican. The commission, voting along party lines, awarded all of the disputed electoral votes to Hayes, who became the nineteenth president of the United States.

The 1876 election rapidly became known as the stolen election, and has the distinction of being the only election in which the losing candidate won a majority of the popular vote. Twelve years later, Grover Cleveland, the incumbent Democratic president, won the popular vote against Republican Benjamin Harrison, 48.6% to 47.8%; however, Harrison won the election with 233 electoral votes to Cleveland's 168.[1]

The next (and most recent) time the plurality winner failed to win the election was in 2000, when Democrat Al Gore won 48.4% of the popular vote against Republican George W. Bush's 47.9%. As in 1876, accusations of voter intimidation and confusing ballots cast doubts on the validity of the voting results in several states—once again including Florida. And once again, the Supreme Court proved decisive in the outcome, effectively awarding all of Florida's electoral votes to Bush, who won 271 electoral votes to Gore's 266.

The fact that the popular vote winner has, on occasion, failed to win the presidential election has led to calls to abandon the Electoral College. But the significance of the Electoral College is magnified by a peculiar feature of American democracy: the office of the president combines, in one person, the functions of head of state and chief executive. In most democracies, these functions are separated. In many, the head of state (the representative of the nation to other countries) is elected directly by the voters. But *no* major democracy chooses its chief executive (the head of government) through a direct popular vote.

The Seats-Votes Curve

A commonly raised objection to the Electoral College is that it favors Republican candidates. We can assess this claim using the *seats-votes curve*, a powerful tool for assessing representative democracies. The seats-votes curve plots the fraction of the popular vote won by a given party against the fraction of the legislative seats won by that party's candidates. Thus we might plot, in any election, the fraction of the total votes for Republican congressional candidates against the fraction of congressional seats won by Republicans. Using a mathematical technique known as *least squares regression*, we can

1. Harrison only served one term. In 1892, he lost—to Grover Cleveland, who earned the distinction of being the only person to have served two non-consecutive terms as U.S. president.

find a formula that relates the percentage of the popular vote won by a party and the percentage of congressional seats won by the party.

We can do something similar with the Electoral College and obtain a formula that relates the percentage of the popular vote won by a party's candidate to the percentage of the electoral vote won by the candidate. Consider the presidential elections from 1876 to 2012. If P is the percentage of the popular vote won by a Republican candidate, and V the percentage of the electoral vote won by the Republican candidate, least squares regression gives us the formula

$$V = 3.077P - 0.9807.$$

For example, the formula predicts that when Republican candidates win $P = 0.50$ (50%) of the popular vote, they will win about $3.077 (0.50) - 0.9807 = 55.78\%$ of the electoral vote.

A similar analysis for Democratic candidates leads to the formula

$$V = 2.9707 P - 0.9211.$$

Using this formula, when Democratic candidates win $P = 50\%$ of the popular vote, they will win about 56.43% of the electoral vote.

Note that this means that if a Democratic and a Republican candidate each receive the same percentage of the popular vote (50% in this case), the Democratic candidate will receive slightly more of the electoral vote (56.43% vs. 55.78%). This means that the Electoral College has a *partisan bias* toward Democratic candidates. Partisan bias reflects a sort of "built-in" advantage to one party.[2]

However, partisan bias is not the whole story. The *responsiveness* or *swing ratio* measures how a change in the fraction of the popular vote won by a party translates into a change in the fraction of the electoral vote won by the party. Mathematically, the responsiveness corresponds to the coefficient of P. For Republicans, winning an extra 1% of the popular vote translates into winning an extra 3.077% of the electoral vote, while for Democrats, the additional 1% leads to an additional 2.9707%. By this measure, the Electoral

---

2. The alert reader might note that if both candidates have exactly 50% of the popular vote, the formulas predict they will have between them more than 100% of the electoral vote. This apparent paradox is caused in part because we are using two different models (the Republican seats-votes curve and the Democratic seats-votes curve) simultaneously, which can lead to conflicting predictions: "A person with one watch knows the time; a person with two watches is never sure."

College is more responsive to Republican voters: "mobilizing the base" is more effective for Republicans than for Democrats.

There are three catches. First, the preceding formulas were based on *all* elections since 1876. The U.S. population has grown more urban, with the traditionally Democratic urban areas growing more rapidly than the traditionally Republican rural regions. Many political scientists find it convenient to use 1948 (the year of the first presidential election following World War II) as a starting point for modern election results. By the same methods as before, the seats-votes curve for Republican candidates in elections since 1948 is

$$V = 3.5359P - 1.17$$

and

$$V = 4.021P - 1.476$$

for Democrats. Using these formulas, we find that the Electoral College is biased toward the Republicans, but more responsive to the Democrats.

There's a second catch: the least squares regression produces a linear relationship between the fraction of the popular vote and the fraction of the electoral vote. However, as early as 1909, political scientists suggested a more complex *power law* for the relationship between the fraction of votes won and the fraction of seats won: In a two-party system, the ratio of the seats won by the two parties is proportional to some power of the ratio of the votes won by the two parties. Based on empirical evidence, most political scientists concluded that the power was the third (hence this observation is also called the *cube law*). Thus, if Party A wins 60% of the vote and Party B wins 40% of the vote, the ratio of the votes won is 60 to 40 (in other words, 3 to 2), but the ratio of the seats won is predicted to be 27 to 8. This would give Party A about 78% of the legislative seats.

The power law can be generalized to form what is called a *bilogit* model of the seats-votes relation. This requires a minor change in our approach, though we can still produce a formula relating the fraction of the popular vote won to the fraction of the electoral vote won, and we still find that the Electoral College is biased toward the Republicans, but more responsive to the Democrats: in other words, it's easier for a Republican candidate to win, but a Democratic candidate benefits more from a gain in votes.

There is one final catch: the preceding analysis is for what is known as the historical seats-votes curve—what has happened in the past. But the number of electoral votes and their geographic distribution has changed over time, as has the political geography of the nation. Thus, even if the Electoral College shows an historical bias toward the Republicans, that bias could be caused by long-term demographic shifts and not the Electoral College itself. The important question is whether the Electoral College affects any given election.

This type of analysis has not been feasible until very recently, as it requires a sophisticated model of elections; enough computational power to simulate them; and accurate data on national sentiment to make the simulations meaningful. It was not until 2013 that such an analysis was performed, by a group of researchers consisting of A. C. Thomas of Carnegie-Mellon University, Andrew Gelman of Columbia University, Gary King of Harvard, and Jonathan Katz of Cal Tech. The group analyzed presidential elections between 1956 and 2004 and concluded that the Electoral College did *not* introduce a statistically significant partisan bias. Thus, it appears that the Electoral College does not give one party an advantage over the other.

One Man, 3.312 Votes?

Even if it did, it's not clear that this would be unconstitutional: the U.S. Supreme Court has consistently ruled that in matters of election politics, it is the rights of the *voters*, not the parties, that are important. If we are to find a viable objection to the Electoral College, we must consider how it affects voters. We might look at how it treats voters in different states.

In most states, the entire electoral vote of the state is given to the candidate who wins the popular vote in that state: this is the "winner-take-all" system. It follows that the president is elected using a weighted voting system. Thus, after considering the weighted voting systems in New York, John Banzhaf considered the Electoral College. As before, Banzhaf used the idea that the importance of a voter is measured by how often the voter turns a losing coalition into a winning one, or vice versa. In terms of the Electoral College, this means that the voter is pivotal in the popular vote of their state, and the state is pivotal in the electoral vote of the nation.

The first is easy to calculate. To begin with, assume that voters are evenly split between the two candidates. We can model the outcome of an election using a random variable based on the assumption that each voter has probability $p=1/2$ of voting for either candidate. Under these assumptions, the probability that a voter is decisive in a general election is inversely proportional to the square root of the number of votes. Thus, since California has about 36 times the population of Montana, the probability that a voter in California is decisive in the California vote is about $\dfrac{1}{\sqrt{36}} = 1/6$ the probability that a voter in Montana is decisive in the Montana vote.

What about the probability that a state is decisive in the electoral vote? This requires a more complicated calculation, but roughly speaking, it is directly proportional to the number of electoral votes cast by the state. Thus California, with 55 electoral votes, is about 18.33 times more likely to be decisive than Montana, with 3 electoral votes.

The combination of the two factors means that the voter in California is $(1/6) \times 18.33 \approx 3.06$ times more likely to be decisive than the voter in Montana. This type of analysis (with more refined estimates on the probability that a state's electoral votes are decisive) led Banzhaf to conclude, in 1968, that voters in California were significantly more likely to be decisive than voters in less populous states, and the title of his article, "One Man, 3.312 Votes" became a new catchphrase for opponents of the Electoral College.[3]

Election Models

Banzhaf's analysis is based on what is known in political science as the *Beck model* of elections. This type of model assumes that every voter has some probability of voting for a candidate, and attempts to predict the election results based on this assumption. There are a number of objections to the Beck model. The most commonly stated is that voters don't flip coins. This is true— and utterly irrelevant. The probability does not model the *process* of voting; it models the *results*, so any real criticism of the Beck model must come from whether or not it yields good results.

The Beck model relies on two parameters: the number of voters, and the probability a voter supports a particular candidate. The first is a matter of public record. The obvious source of the second is pre-election polls. For example, in the days leading up to the 2012 election, polls in Florida showed Republican candidate Mitt Romney trailing Democratic incumbent Barack

---

3. Ironically, Banzhaf's title references the idea that a person with k votes has k times as much political power, a notion that Banzhaf himself was instrumental in refuting. But "One Man, 3.312 As Much Chance of Altering a Presidential Election Outcome" is a more awkward title.

Obama, 47% to 49%. Using the Beck model, we would predict that Romney would win between 46.9% and 47.1% of the Florida vote, and Obama would win between 48.9% and 49.1%. The actual election results were 49.1% for Romney and 50.0% for Obama. While this seems to be a small percentage difference, the Beck model predicts that Obama would have had a better chance of winning the lottery—fifty times in a row—than of winning 50.0% of the Florida vote. Does this mean that election irregularities in Florida have once again altered the course of American history?

Perhaps—but it is more reasonable to suppose that the Beck model is flawed. As we saw in the section on the census, the sample mean will generally differ from the population mean. In this case, of a sample of voters, 49% chose Obama. But the fraction of voters in the population who would vote for Obama will in general be different. This difference is usually reported as the margin of error: a poll might have indicated that 49% ± 4% of the voters supported Obama, and 47% ± 4% supported Romney.

As with the census, this means that the observations support a conclusion that somewhere between 45% and 53% of the voters supported Obama, and between 43% and 51% supported Romney. Within these possibilities, anything could happen: the vote might split 49% Romney and 45% Obama, giving victory to the Republicans; or 48% Obama, 47% Romney, giving victory to the Democrats. A cautious commentator might call the state a toss-up.

However, the prediction of Romney winning between 45% and 53% of the vote is nothing more than a sound bite. We might reason as follows. The polls support the claim that between 45% and 53% of the voters will cast a vote for Obama. Somewhere in this interval is the correct probability that a voter will support Romney, and if we select the right probability, we can make an accurate prediction of who will win the state.

We might pick 47%, which would give us the (incorrect) prediction that Obama would win between 46.9% and 47.1% of the vote. Or we might pick 51%, which would give us another incorrect prediction that Obama would win between 50.9% and 51.1% of the vote. Or we might pick 50.0%, which would give us a correct prediction that Obama would win between 49.9% and 50.1% of the vote.

Of course, we don't know in advance which probability to choose. Fortunately, mathematics allows us to choose all of them. This leads to what is known as the *Good-Mayer model* of elections. We can proceed as follows.

First, we'll assume the 4% margin of error corresponds to a 95% confidence interval, which is fairly common practice among pollsters.

Consider Romney's poll data, which supports the assumption that between 43% and 51% of the voters will support him. For now, assume that the actual percentage that supports Romney will be either 43%, 47%, or 51%.

Suppose that 43% of the voters support Romney. We can compute the probability that this will occur: it is about 15.73%. If this occurs, then Obama need only win more than 43% of the vote to win the state of Florida. The probability that this happens will be 99.87%. On the other hand, suppose that 47% of the voters support Romney. As before, we can compute the probability that this happens: it is about 68.27%. Again, Obama needs to win more than 47% of the vote to win the state: this will occur with probability 84.13%. Finally, the probability that Romney wins 51% of the vote will be 15.73%, and if this happens, Obama has no chance of winning the state of Florida.

We can multiply and add these probabilities to find the probability that Obama wins the state of Florida; it will be

$$(0.1573)\,(0.9987) + (0.6827)\,(0.8413) + (0.1573)\,(0) = 0.7315$$

While a Romney win is possible, it isn't very likely. A more sophisticated computation requires calculus, but uses essentially the same logic as above; if we do this, we find the probability that Obama wins Florida to be about 86%.

If we repeat this type of computation for all 50 states and the District of Columbia, we can find the probability that a candidate will win the national election. Statistician Nate Silver performed this sort of analysis prior to the 2008 presidential election, and correctly predicted the outcome in the District of Columbia and 49 out of 50 states; he also correctly predicted all senatorial elections that year, using the same methods.

When he performed a similar analysis prior to the 2012 election, Silver concluded that Barack Obama would win re-election easily. Conservative political commentators offered their own predictions. Karl Rove predicted a narrow victory, with 279 electoral votes for Romney, while Michael Barone gleefully predicted a Romney landslide, with 315 votes. Since Silver's predictions flew in the face of partisan wishful thinking, many derided his statistics and attacked his methodology as flawed. Ironically, they were able to do so because Silver, like all reputable mathematicians and scientists, explained his methods. So when he rejected a poll as too inaccurate to be meaningful, he made clear that he did so and provided ample evidence supporting his decision to reject the poll.

The results speak for themselves: Silver successfully predicted all 50 state votes and the correct electoral total. More dramatically, Silver not only predicted which candidate would win the state, but also predicted the final vote tallies to a high degree of accuracy. Silver himself attributes his successful predictions to "bothering to obey the laws of probability theory." These laws of probability theory (not to mention Silver's success) support the Good-Mayer model as a reliable means of predicting election outcomes.

What if we use the Good-Mayer model to find the probability a voter is decisive in a presidential election? The probability that a state is decisive is unchanged, and still roughly proportional to the number of electoral votes held by the state: California is about 18.33 times as likely to be the decisive state as Montana. However, the probability a voter in a state is decisive decreases considerably: rather than being inversely proportional to the square root of the number of voters (which it is under the Beck model), it is inversely proportional to the *number* of voters. A California voter is about 1/36 as likely to be decisive as a Montana voter. Together, these two factors imply that a voter in California is $(1/36) \times 18.33 \approx 0.51$ times as likely to be decisive as a voter in Montana. Hence, a voter in California is *less* likely to be decisive than the voter in Montana.

Defending the Electoral College: Voter Turnout

The fact that the Beck and Good-Mayer models lead to opposite conclusions regarding the probability that a voter is decisive suggests that this type of attack on the Electoral College might not be effective. In fact, there is a more fundamental problem: valuing voters in different states is constitutionally acceptable.

Consider the Senate. Since every senator casts one vote, every senator is equally likely to be decisive in any given vote. But since the entire population of a state elects its senators, a voter in a more populous state (regardless of which model we use) is less likely to be decisive in the election of their senator than a voter in a less populous state. Hence a voter in more populous states is less likely to be the decisive voter in the election of the decisive senator. Not only is this disparity built in to the Constitution; Article V effectively prohibits any amendment from changing it: "No state, without its consent, shall be deprived of its equal suffrage in the Senate."

In any case, pointing out flaws in the Electoral College is disingenuous if we do not, at the same time, carefully examine the alternatives. The most obvious replacement is a direct popular vote. This is intuitively appealing, since the contentious elections of 1876, 1888, and 2000 have all centered around a mismatch between the popular vote winner and the electoral vote winner. This is the basis of the National Popular Vote Plan, which would have each state assign its electoral vote on the basis of the national popular vote.

As before, we might want to consider the probability that a voter is decisive in the two systems. We have two models of elections: the Beck model and the Good-Mayer model. The latter has the virtue of predicting actual election results with a high degree of accuracy; it is tempting to discard the less accurate Beck model entirely and use the Good-Mayer model exclusively.

However, there is a crucial difference between the two models. As used above, the Beck model assumes that voters *are* evenly split between two candidates. In contrast, the Good-Mayer model assumes that voters *appear* evenly split between the two candidates. In short, the Beck model assumes something about reality, while the Good-Mayer model assumes something about our perception of reality.

Both viewpoints are important. To understand why, consider the following: We can easily construct a model that predicts, to a high degree of accuracy, the number of jobs requiring mathematics PhD's that are held by women. On the other hand, we can construct a model based on the fact that women constitute about 50% of the population, and predict that about 50% of the jobs requiring mathematics PhD's would be held by women.

The second model makes predictions that are so inaccurate as to be useless. But it does not follow that the second model is irrelevant! In fact, we might use it as a guide to what *should* be. In the same way, we can use the Beck model to provide useful insight into how a democratic system *should* work.

With this in mind, let us consider what happens when we use the Beck model to compare the probability a voter is decisive in a popular vote system, to the probability a voter is decisive in a system like the Electoral College. Such an analysis was done by Alan Natapoff of the Massachusetts Institute of Technology. To understand Natapoff's analysis, let's begin with a simple system: two candidates and 125 voters divided into five geographic districts with 25 voters apiece. There are two cases to consider: first, a popular vote system, where the winner will be the candidate with more total votes, regardless of which districts those votes came from. Second, a districted vote system, where the winner will be the candidate who wins more districts, where a district is won by winning more votes in that district.

In a popular vote system, a voter will be decisive if the remaining $125 - 1 = 124$ voters are split 62 to 62. If each voter is equally likely to vote for either candidate, then the probability that the remaining voters will split 62 to 62 is about 7%: roughly 1 in 14.

In the districted system, a voter will be decisive when the remaining $25 - 1 = 24$ voters in their *district* split 12 to 12, *and* the other $5 - 1 = 4$ districts split 2 to 2. The first probability is about 16%, and the second probability is 38%. The probability that a voter is decisive in their district *and their* district is decisive in the election will be 16% of 38%, which is about 6%, or 1 in 16. It follows that a voter is less likely to be decisive in a districted system than in a popular vote system.

Natapoff's analysis goes further and considers the case where the voters are *not* evenly split between the two candidates. For example, suppose one candidate is supported by 60% of the electorate and the other by 40%. In this case, the probability a voter is decisive in a popular vote system is about 1 in 175. However, the probability that a voter is decisive in a districted vote system is higher: about 1 in 100.

With these probabilities in mind, let's consider voting from a game theoretic standpoint. There are two strategies: vote, or don't vote. The act of casting a vote incurs some cost: for example, the waiting time at the polling place. At the same time, there are two possible outcomes: your candidate wins, or your candidate loses. We might associate the first outcome with a gain and the second with a loss.

Suppose we knew, in advance, that our candidate was going to win without our vote. Then whether we voted or not, we would obtain the same payoff, namely that our candidate won; however, if we voted, we would have to subtract the cost of voting from that payoff. This means that not voting gives us a higher payoff. Likewise, suppose we knew, in advance, that our candidate was going to lose even with our vote. Then, whether we voted or not, our loss would be the same, namely that our candidate lost; however, if we voted, this loss would be compounded by the cost of voting. Not voting helps reduce the loss. In both cases, the rational actor model would lead us to conclude that it is better *not* to vote: we will either gain a higher payoff, or incur a smaller loss. Put succinctly: dictatorships have low voter turnout.

What if the candidate is not certain to win and not certain to lose? In this case, our choice of strategies will depend on the probability that ours is the decisive vote (since if our vote is not decisive, then our candidate is either certain to win, or certain to lose, and our vote will not affect the outcome).

Again, suppose that we are *certain* to cast the decisive vote. Then there is no question that we should vote, since if we do, our candidate will win; otherwise, our candidate will lose. Likewise, suppose there is *no* chance we will cast the decisive vote. Then there is no point in voting: our candidate will win or lose without our support, so we might as well avoid the lines at the polling place. It follows that somewhere between *no* chance of casting the decisive vote, and absolute *certainty* of casting the decisive vote, our strategy will switch between not voting and voting.

Consider our example, with 125 voters. If the voters are evenly split between two candidates, a voter in a popular vote system would have a greater probability of being decisive than a voter in a districted system. It follows that the voters in the former would be more likely to vote than the voters in the latter, and voter turnout will be higher. However, if the voters are split 60–40 between the two candidates, voters in the districted system are more likely to be decisive than voters in the popular vote system. By a similar argument, voter turnout in the districted system will be higher.

In our system, somewhere between the situation where one candidate has a 0% lead and the situation where one candidate has a 20% lead, voters will be more likely to vote in the districted system than in the popular vote system. The turning point occurs at the *critical difference*. In our example, the critical difference is about 10%, so if one candidate has a 10% lead over the other, voters are equally likely to be decisive in both systems; but if one candidate leads by more than 10%, voters are more likely to be decisive in the districted system.

The critical difference falls as the number of voters increases. For example, if there are one million voters split into five districts, the critical difference is only 0.2%. Most recent U.S. elections give one candidate at least this much of a lead over their opponent. It follows that the practice of dividing the electorate into districts increases the probability a voter is decisive, and consequently voter turnout is higher than it would be otherwise.

Of course, these results are based on using the Beck model, which is an admittedly terrible model for how real elections work. But as it turns out, we reach similar conclusions using the Good-Mayer model of elections. In general, we would predict that voter turnout in a districted system will be higher than it would be otherwise.

Defending the Electoral College: Special-Interest Groups

It would be nice to have empirical evidence supporting the conclusion that voter turnout is lower in popular vote systems. However, disentangling the data is impossible: presidential elections, decided by districted systems, present voters with choices and consequences very different from local elections, decided by popular vote. A meaningful comparison could only be made if we ran the same election twice—once using a districted system and a second time using a popular vote system. Thus we might discount higher voter turnout as a reason to support the Electoral College.

There is a second point in favor of the Electoral College. Suppose that *most* of the electorate is evenly split between two candidates, but there is a special-interest group who will throw its support behind one of two candidates. How important are such groups?

As before, we might apply the Beck model to a simple electorate with 125 voters. Suppose that five of the voters (4% of the electorate) belong to a special-interest group. In a popular vote system, a majority of 63 of the 125 votes is necessary to win. The special-interest group will be decisive if a candidate wins between 58 and 62 of the remaining 120 votes. This will occur about 35% of the time.

On the other hand, if the 125 voters are divided into five equipopulous districts, a number of scenarios present themselves. There are two extremes: first, the *sectional case*, where all members of the special-interest group live in same district; second, the *national case*, where all members of the special-interest group are evenly distributed among all the districts.

In the sectional case, the five members of the special-interest group would constitute a substantial fraction of the district population, so they would be very likely to be decisive in their district; indeed, we find that they will be decisive about 74% of the time! However, they will only be decisive in the overall election if their district is decisive, which requires that the remaining four districts split 2 to 2. This reduces the probability that the special-interest group is decisive to about 28%.

In the national case, the five members would be dispersed among the five districts. This time, the special-interest group will be decisive in a number of cases: for example, if they are decisive in one district and the other two districts split 2 to 2; if they are decisive in two districts and the other districts split 2 to 1 or 1 to 2; and so on. The probability that this occurs will be about 29%. In the districts where the special-interest group's vote was not decisive, the candidate must win a majority *without* their vote.

What shall we make of these results? First, in both cases, the special-interest group is less likely to be decisive in the districted system than in the popular vote system. Moreover, the reduction is *greater* in the sectional case than in the national case. The disparity grows with the size of the special-interest group. A regional special-interest group with 15 voters would always be decisive in their district (since they command a majority by themselves), but would only be decisive if the remaining districts split 2 to 2: this occurs about 38% of the time. A national special-interest group with 15 voters (with 3 members in each of 5 districts) would be decisive 69% of the time. But in a popular vote system, this special-interest group, which makes up 12% of the population, would be decisive about 85% of the time! In other words, districted systems dilute the power of special-interest groups.

In fact, we can go further. Consider two equally large special-interest groups—one "national," with adherents in all districts, and the other "sectional," with adherents in just one district. As before, we will assume the special-interest group to have 5 members, and the electorate to have a total of 125 voters.

Now consider two candidate seeking support from these groups. Suppose that one candidate receives the endorsement of the sectional special-interest group, while their opponent receives the endorsement of the national special-interest group. If all other voters remain equally divided between the two candidates, we find that the candidate endorsed by the *national* special-interest group has a 50.6% chance of winning the election, while the candidate endorsed by the *sectional* special-interest group has a 49.4% chance of winning the election. Thus, districted systems reward candidates supported by national special-interest groups, at the expense of candidates who appeal to purely sectional interests.

The Electoral College Is the Worst Form of Democracy, Except for All the Rest

Ironically, one of the more common arguments *against* the Electoral College refers to this very feature: namely, the argument that a candidate need only win 11 states to win the presidency. This is true, but the states represent the west (California, with 55 votes), south (Texas, Florida, North Carolina, Georgia, and Virginia, with 104 votes), northeast (New York, New Jersey, with 43 votes), and midwest (Pennsylvania, Illinois, Ohio, and Michigan, with 74 votes). So, even though a candidate could win the election without winning the popular vote, he or she would still have widespread national support.

The Electoral College, as it is currently implemented, differs from the districted systems analyzed above. This is because the districts are themselves aggregated into larger units (namely, the states) and in most states, the entire electoral vote of the state is given to the candidate who wins the popular vote in the state: the winner-take-all rule. Moreover, each state has a "senatorial bonus" of two electoral votes that are independent of the number of actual voters in the state. This is, in fact, the primary source of the problem that a candidate can win the election without winning the popular vote.

What if we prohibit the winner-take-all allocation of electoral votes? This would retain all the desirable features of districted systems, while removing an artificial and unnecessary feature that magnifies the inequities produced by differing state populations. Two states already assign electoral votes in this fashion (Maine and Nebraska, with the senatorial votes going to the winner of the popular vote in the state).

However, the Maine and Nebraska plans introduce another problem: the congressional electoral votes are assigned according to the vote in the district. Thus, they bring to the Electoral College all the risks from unlimited partisan gerrymandering. In fact, the 2013 study that showed the lack of bias in the Electoral College under the winner-take-all rule also showed that if electoral votes were assigned on a district-by-district basis, there would be a significant partisan effect in favor of Republican candidates, exacerbated when the two senatorial votes are given to the winner of the state popular vote.

Thus, whatever flaws the Electoral College has, the alternatives are no better. A direct popular vote sounds good in principle, but districted systems reward candidates who support (and are supported by) national concerns and offer protection from regional special-interest groups. Meanwhile, assigning the congressional electoral votes on a district-by-district basis risks introducing a significant partisan bias due to the persistence of partisan gerrymandering. Winston Churchill once opined that democracy is the worst form of government, except for all the rest; we are in a similar position with respect to the Electoral College.

Part II / The Bill of Rights

Stop and Frisk

The right of the people to be secure in their persons, houses, papers, and effects, against unreasonable searches and seizures, shall not be violated.   —U.S. CONSTITUTION, AMENDMENT IV

In the early morning hours of September 4, 1992, police in Oneonta, New York, responded to calls of a break-in at a house where a 77-year-old woman was visiting. When they arrived, she reported fighting off an assailant. Although she was unable to get a clear view of his face, she described him as black, based on her view of his hand and forearm; young, because of the speed with which he crossed her room; and wielding a knife. A blood trail suggested that the attacker had cut himself on the hand or forearm with the knife.

Oneonta is a largely white city in upstate New York; the largest contingent of blacks was at the nearby State University of New York at Oneonta. Oneonta police requested and obtained from the university the names and addresses of every black student; over the next few days, police visited their residences and questioned them, inspecting their hands for cuts. They also stopped and inspected the hands of nonwhites on the streets of Oneonta. And even though the suspect was described as being male, they also stopped three black women, including Sheryl Champen, then assistant director of admissions at the university. Despite these efforts, police failed to find a suspect.

Although the victim described her assailant as a black male, the constitutional question is: Was it reasonable to search young black males (and three black women) *because* of their race? Champen joined others in a class-action suit filed against the city of Oneonta. In *Brown v. City of Oneonta* (2000), the Second Circuit ruled that, *in general*, the description of the perpetrator justified the stopping of black males, though *in particular* their actions against Champen and several others were excessive. In 2006, Champen and one of the students, Ricky Brown, won a judgment against the city of Oneonta.

Terry Frisks and Traffic Stops

There are two primary approaches to law enforcement. In the *warrant model*, law enforcement becomes involved after a crime has been committed, with a goal of tracking down the perpetrators. In contrast, the *patrol model* focuses on preventing crime in the first place by having officers present in the community. Obviously, both strategies are needed. But since most people would prefer that crimes didn't happen in the first place, the patrol model should be the mainstay of law enforcement strategy.

The mere presence of law enforcement officers is insufficient to prevent all crime; they must be able to take preventive actions as well. For example, in 1963 Cleveland plainclothes police officer Martin McFadden observed suspicious behavior by John Terry and Richard Chilton. McFadden approached them, identified himself as a police officer, and searched the men; he found a pistol in Terry's jacket and a revolver in Chilton's. Both men were subsequently charged with carrying concealed weapons. Terry appealed, arguing that McFaddden's search was unreasonable. In *Terry v. Ohio* (1968), the U.S. Supreme Court ruled that reasonable suspicion that the subject has a dangerous weapon, not probable cause, is sufficient grounds for a frisk.

There was one important factor in *Terry* that the Supreme Court did not address: McFadden was white, while Terry and Chilton were black. The natural question was whether McFadden interpreted the behavior of Terry and Chilton as suspicious because he was white and they were black. Allegations of racist behavior by police led to riots in Harlem (July 16–22, 1964), Philadelphia (August 28–30, 1964), Watts (August 11–17, 1965), Detroit (July 23–27, 1967), and other cities. As a result, the patrol model fell out of favor and during the 1970s the warrant model became more common.

The rise of the warrant model coincided with the rise of the term *individualized suspicion* with regard to 4th Amendment cases. However, "individualized" is a meaningless modifier. To see why this is so, consider the situation in *Brown v. City of Oneonta*. We might object to police actions on the grounds that none of those stopped were suspicious for any reason *other* than their race; thus the authorities did not have *individualized* suspicion. But suppose that the police were correct, and that the perpetrator was one of the black males in Oneonta. Furthermore, suppose that there was only one black male in Oneonta at the time. It seems hard to deny that the police would have, in such a circumstance, "individualized suspicion."

But what if there were two black males in Oneonta? By the doctrine of "individualized suspicion," the police would not be able to interview either one. This seems an unreasonable restriction of law enforcement's ability to pursue a suspect. At the same time, if most of the city's population were black, we might hold it unreasonable to search all of them. We might ask: At what point does the fact that a person matches some or all of the characteristics of a perpetrator provide authorities with individualized suspicion?

To answer this question, let's consider the probability that a person is the perpetrator. In 1992, the black male population of Oneonta was about 150. Assuming the perpetrator was one of these men, the probability that a randomly stopped black male was the perpetrator was 1 in 150. Our rejection or acceptance of police actions should be based on this probability: is it high enough for a search and seizure to be reasonable?

We might apply the same logic to the patrol model: a person's characteristics and behaviors can be translated into the probability that he or she is engaged or about to be engaged in a criminal act. Above some threshold probability, a search and seizure becomes reasonable. The constitutional question is identifying a threshold probability of wrongdoing. Mathematics can then be used to determine whether the searches that are conducted meet this threshold probability.

A lower limit may be gleaned from *Michigan Department of State Police v. Sitz* (1990), which concerned the constitutionality of highway checkpoints for sobriety. Nothing, other than the fact that a vehicle was being driven through a checkpoint, determined whether a driver would be questioned. The Supreme Court ruled that since the questioning was minimally invasive and inconveniencing, it satisfied the 4th Amendment's requirement for reasonableness. Since about 1.6% of the drivers were found to be intoxicated, the Court's decision suggests a threshold probability of 1.6%.

In contrast, the Court ruled that a similar checkpoint to search for illegal drugs was *not* constitutional in *City of Indianapolis v. Edmond* (2000). To make matters even more confusing, the checkpoints in *Indianapolis v. Edmond* found contraband in about 5% of the searches. Thus, while *Michigan v. Sitz* suggested that a 1.6% threshold probability was high enough to justify a search when checking for sobriety, *Indianapolis v. Edmond* suggests that a 5% threshold probability was too low to justify a search when checking for narcotics.

It's possible to reconcile the two cases by noting that a check for sobriety is a quick, minimally invasive search, while a check for smuggled contraband requires a more detailed search. We might set the threshold probability of wrongdoing higher when exceeding it would draw down a more inconveniencing and more invasive search. Additionally, a drunk driver is an immediate danger, while someone smuggling drugs is not. In such a situation, we might want to allow a reduced threshold probability of wrongdoing. The Court chose to focus on the fact that the sobriety checkpoints were directed against a specific violation of the law, while narcotics checkpoints were motivated by a "general interest in crime control."

Chief Justice Rehnquist, one of the three dissenters in *Indianapolis*, rejected the majority argument that a "general interest in crime control" should be controlling. Instead, Rehnquist argued that the reasonableness of the stops, and the fact that they were conducted in a neutral manner, should be decisive: "It is the objective effect of the State's actions on the privacy of the individuals that animates the Fourth Amendment" (531 U.S. 32, p. 52).

The Cold Equations

Rehnquist's viewpoint implies that as long as a stop is *neutral* and reasonable, it should be allowable. Checkpoints represent a particularly easy situation: either all vehicles passing through are stopped (as in *Michigan*), or a predetermined selection of vehicles are stopped (as in *Indianapolis*). But in general, officers have more discretion over who they stop. This leads to accusations that a person is stopped for "driving/loitering while black."

In May 1992, Robert Wilkins and his family were returning from a funeral in a rental car when Maryland police stopped them for speeding. Subsequently their vehicle was searched for illegal drugs (none were found). Wilkins, a Harvard-educated lawyer (and now a federal judge) filed suit, charging that the stop was based on a racial profile. Maryland police settled after a document surfaced that contained an explicit profile targeting African Americans; part of the settlement agreement required the police to maintain detailed records of traffic stops in order to evaluate whether they had changed their ways. Two years later, the data were analyzed by John Lamberth of Temple University. Lamberth found that while blacks constituted 17% of the drivers, 72% of those stopped and searched were black. This gross disparity suggests that police are targeting black drivers at a much higher rate than white drivers.

The standard response to this sort of accusation is that minorities are more likely to draw police scrutiny because minorities commit crimes at a higher rate. In connection with a case we'll discuss later, Judge Shira Scheindlin of the U.S. District Court for the Southern District of New York, gave the standard response to this defense: "Many police practices may be useful for fighting crime—preventive detention or coerced confessions, for example— but because they are unconstitutional they cannot be used, no matter how effective" (08 Civ. 1034 (SAS), p. 5). Unfortunately, this response pits two abstractions against each other: the need for equal treatment before the law, and the need for enforcement of the law. When abstractions battle, victory tends to go to the side that shouts the loudest.

A better approach is to compare the expected value of various strategies and adopt the one with the highest expected value: in other words, to follow the optimal strategy. The *rational actor model* assumes that people will choose the optimal strategy. This appears to be an unduly optimistic viewpoint, and in the short run people can and do choose suboptimal strategies. However, in the long run, those who choose suboptimal strategies will do worse than those who follow optimal strategies.

This is the alleged basis for libertarianism and social Darwinism, which hold that people should be allowed to follow suboptimal strategies and live with the consequences. However, these philosophies miss a crucial point: Nothing prevents a society from committing suicide by making suboptimal choices—and the laws of nature don't care whether a society survives or not. Since those who live in the society might care, it follows that identifying the optimal choice *and following it* is an important task for policymakers.

We might consider several policy choices. If one strategy is more effective, we should choose it; if the two strategies are equally effective (and more effective than any of the others), it makes no difference which one we follow.

It's important to realize that in this computation, we must ignore *any* factor that cannot be quantified. In particular, until we can assign a specific value to abstractions like "justice" and "liberty," they can play no role in the decision-making process.

This refusal to incorporate unquantifiable abstractions leaves many uneasy, since the ultimate quantifier in our society is monetary value. As a result, the mathematical viewpoint leads to the viewpoint that everything can be reduced to an amount in dollars and cents: as the saying goes, money is a way to keep score. But it's actually the best way to analyze different strategies, for two important reasons. First, it forces us to consider the actual costs and benefits of a strategy. And second, it forces us to consider who pays the costs and who reaps the benefits.

For example, motivated by U.S. foreign policy and a belief in a divine mission, a group of extremists orchestrated an attack on U.S. soil that left thousands dead and a nation in shock, fearful that a similar event would occur in the near future. The events of that day—December 7, 1941—led President Franklin Roosevelt to issue Executive Order 9066 on February 19, 1942.

Executive Order 9066 ordered the relocation of all civilians of Japanese ancestry in the western states to concentration camps, ostensibly to guard against acts of sabotage perpetrated by Japanese sleeper agents in the United States. While some of these civilians were born in Japan, many of them had been born in the United States and so, by the 14th Amendment, were automatically U.S. citizens. Executive Order 9066 has been called the most egregious violation of civil rights in the history of the United States, but faced almost no opposition from the Supreme Court, politicians, or activists. Ralph Lawrence Carr, then-governor of Colorado, was one of a very small number of elected officials to object to this action, a heroic stance that would cost him re-election.

Such internment did not occur after the terrorist attacks of September 11, 2001. But the only thing that prevented it was our self-restraint. Even so, a rabid minority (their isolated voices magnified via the Internet) called for such actions, and others, while rejecting the notion of universal detention of persons of Middle Eastern ancestry, accept increased scrutiny of such persons as a reasonable tradeoff between civil liberties and security.

But what do the cold equations tell us? For simplicity, consider just two strategies: first, incarcerate the roughly 1.5 million Americans of Middle Eastern ancestry in detention camps. Alternatively, do nothing. The first strategy has several clearly quantifiable costs. First, the annual cost of housing one prisoner in a federal facility is about $20,000. Next, wrongfully incarcerated federal prisoners are entitled by law to up to $50,000 per year of incarceration. If we assume that most persons of Middle Eastern ancestry are not terrorists, then the cost of interning 1.5 million persons will be $105 billion each year.

What if we do nothing? The results of this strategy are more difficult to evaluate, but suppose we assume that among the 1.5 million persons of Middle Eastern ancestry there is a group that will successfully perpetrate a 9/11-scale attack during the year.[1]

In order to make a meaningful comparison between the two strategies, we must consider only the same type of costs. The costs of the first strategy would be paid by federal, state, and local governments, but we have ignored any personal losses incurred by those affected, such as the loss of business and educational opportunities. Hence when we evaluate the cost of a terrorist attack, we should consider only the costs paid by federal, state, and local governments, but no personal losses incurred by those affected. Therefore, of the property damage caused by the 9/11 attacks, only the damage to the Pentagon and municipal property, as well as the cost of cleaning up the debris, should be included: this totaled about $9 billion.

There were additional costs to federal, state, and local governments as well, in the form of tax revenues lost due to deaths and business interruptions. To exaggerate the cost of a terrorist attack, let us be very generous in determining these amounts. First, the roughly 3,000 dead represent lost tax revenues from taxes paid during their working lives. If we assume they would have worked for another 20 years and paid $50,000 in taxes each year, then a 9/11-style attack would lead to another $3 billion loss. We should amortize this loss over the lifetime of the taxpayer. However, this would reduce the "present value" of the loss, and since our goal is to be as generous as possible in computing the cost, we will consider the unamortized value.

---

1. Law enforcement stops many terrorist attacks, but bad luck and poor planning also play a part in failure rates. The first attack on the World Trade Center buildings, on February 26, 1993, was intended to knock down both towers and cause 10,000 casualties. It killed six people and destroyed a parking garage. On December 25, 2009, a terrorist tried to detonate explosives smuggled onto an airplane in his underwear, but only managed to set his pants on fire. On May 1, 2010, another terrorist tried to explode a car bomb in Times Square, but only managed to set the car on fire.

Business interruption is more difficult to evaluate. We might consider the Gross Domestic Product (GDP). The inflation-adjusted GDP grew only 0.35% between 2000 and 2001, compared with a historical average of about 3%. If we attribute *all* of the difference to the events of September 11 (when the GDP actually fell during the first and second quarters, indicating the U.S. economy was in a recession), this would translate to a reduction in GDP of about $300 billion. Again, we should only consider the costs to federal, state, and local governments, which would come from a loss in tax revenue: at a combined corporate tax rate of 30%, this translates into $90 billion in lost revenues. Thus, using the actual governmental costs and some very generous assumptions over the economic losses, the terrorist attacks incurred a cost of about $102 billion.

Now we can objectively analyze our two strategies. The strategy "Detain all persons of Middle-Eastern ancestry" incurs a cost of $105 billion over the next year. The strategy "Allow a 9/11-style terrorist attack to occur during the next year" incurs a cost of $102 billion over the next year. We need not resort to abstractions like civil liberties or constitutional freedoms: there are three billion reasons why the detention strategy is a foolish one.

Race-Based Policing

What about race-based policing as a strategy? There are two scenarios to consider. First, there might be no real difference in the offending rates: in other words, blacks and whites might *in fact* break the law at identical rates. Thus in Wilkins's case, Lamberth showed that blacks and whites violated traffic laws at roughly equal frequencies.

Although it's "obvious" that in such a case, there is no justification for stopping blacks more often than whites, we cannot base our conclusions on what is "obvious," as this leads us back to trading abstractions. We must evaluate our two policies: focusing scrutiny on members of one race, or not.

Suppose 10% of blacks and 10% of whites are smuggling drugs (the rationale for the police stops in Maryland). Using Lamberth's data, about 17% of drivers are blacks. Suppose that, over a period of time, the police could stop 1,000 vehicles. Let us assess two strategies: either the stops are done without regard to the race of the driver, or all stops are directed against black drivers.

In the first case, since blacks make up 17% of the drivers, the police would stop 170 black-driven vehicles and 830 white-driven vehicles. Our assumption that blacks and whites offend at the same rate leads us to conclude that this strategy would identify a total of 100 offenders: 17 blacks and 83 whites.

On the other hand, suppose the police focused only on black drivers. They would stop 1,000 black-driven vehicles, 100 of which would be carrying drugs. Again, they would net a total of 100 offenders. Since this strategy has the same yield as the other, both are equally efficient: the cold equations give us no reason to prefer one over the other.

We might be tempted to justify the race-neutral policy over the race-based policy using an abstraction like "fairness." But we must reject this approach, for if we use abstractions to decide between strategies of equal effectiveness, we run into the same problem as before: different people place different values on abstractions. In this case, the counterpoint to fairness is *convenience*: it's easier to judge someone based on the color of their skin.

Instead, we might look more deeply at our analysis. In particular, we should consider what happens when our policy is in effect for some period of time. To begin with, suppose that we have 10,000 motorists, 1,700 of whom are black and 8,300 who are white; for convenience, we may assume that these are all the motorists that frequent a particular section of road. Suppose that 10% of either black or white motorists are carrying contraband. Thus, there are 170 black and 830 white motorists carrying contraband.

Now consider a race-based policy that stops 1,000 black motorists. This group will include 100 of those carrying contraband. Presumably, this group will be removed from the general population. But this means that, of the remaining 1,600 black motorists, only 70 carry contraband; meanwhile, 830 of the 8,300 white motorists carry contraband. The offense rate for blacks is now $70/1600 \approx 4.375\%$, while the offense rate for white motorists remains at 10%.

Suppose we continue our race-based policy, and again stop 1,000 black motorists. Since only 4.375% of them carry contraband, our yield will be about 44 smugglers. In contrast, if our 1,000 stops were directed against white-driven vehicles, our yield would be much higher: 100 smugglers. As a result, the focus on black-driven vehicles is no longer the optimal strategy. Thus, while the race-based policy may be as effective as the race-neutral policy *in the beginning*, it will quickly become suboptimal.

Of course, this assumes that no additional smugglers enter the fray. One might argue that since police activity removed 100 smugglers, they would be replaced by 100 new smugglers. One possibility is that the 100 new smugglers are all black. It should be obvious that, if we continue our race-based policing efforts, over time, *all* blacks would become offenders. In criminology, this is known as a *ratchet effect*: law enforcement policies create a situation where a subpopulation has a higher offending rate.

Actually, it's unlikely that all blacks would become smugglers, for there is a secondary effect at work. Suppose that police were targeting blacks exclusively. A distributor hoping to smuggle drugs does not benefit by having their shipments intercepted. Thus if black-driven vehicles are being searched while white-driven vehicles are not, the distributor will shift to using whites to carry drugs, secure in the knowledge that their vehicles won't be searched. Because whites would be carrying drugs at a higher rate, focusing on black-driven vehicles will once again be a strategy with a lower yield than a strategy not based on racial profiling.

Simpson's Paradox

It should be clear that our initial strategy, based on an assumption of equal offense rates among the groups, will quickly produce a situation where the offending rates are different. Thus we should identify the optimal strategy when there is a real difference in offending rates.

Finding the real rate at which different groups break the law is very challenging. In particular, most of the evidence that minorities offend at a higher rate is based on victim reports and incarceration rates. But the race of the perpetrator is unknown in many crimes, like vandalism or burglary, and in others the victim might not be able to make an accurate identification. Meanwhile, the incarceration rate is only partially determined by the rate at which minorities commit crimes; it also depends on the rate at which they are captured, tried, and convicted.

Even if we do find a real difference in offending rate, we might be unable to translate that difference into an effective law enforcement strategy. Consider table 7, which gives the breakdown of the residents of a fictional city. City-wide, we see that 25% of the population is black, and that 17% of the black population carries drugs. Moreover, we see that $4000/6000 = 67\%$ of those who carry drugs are black.

Now suppose that, as part of a drug interdiction program, police stop and frisk a certain number of people. There are three possible strategies. First, since blacks make up 25% of the population, we might direct 25% of our stops

Table 7. Offending Rates (Hypothetical City, Aggregated Data)

	Black	White	Total
Carries drugs	4,000	2,000	6,000
Does not carry drugs	20,000	70,000	90,000
Totals	24,000	72,000	96,000
Offending Rate	17%	3%	

against blacks and 75% against whites. Next, since blacks make up two-thirds of our offenders, we might direct two-thirds of our stops against blacks and one-third against whites. Finally, since blacks have a *higher* offending rate than whites, we might direct *all* of our stops against blacks.

Since each search takes a certain amount of time, it follows that only a certain number of persons can be stopped and frisked. Suppose that, over some period of time, 1,000 persons are subject to this measure. How do our three strategies fare?

First, a strategy based on the fact that 25% of the *population* is black would stop 250 blacks and 750 whites. Since 17% of blacks carry contraband, this strategy would yield about 43 blacks carrying drugs. Likewise, since 3% of the whites carry contraband, this strategy would yield about 23 white offenders. Thus, basing the strategy on the racial composition of the population would yield $43 + 23 = 66$ offenders.

Next, a strategy based on the fact that 67% of the *offenders* are black would stop 670 blacks and 330 whites. Again, 17% of the blacks and 3% of the whites would be carrying contraband, so we would have a total yield of about 124 offenders. Finally, if we directed all our stops against blacks, we would stop 1,000 blacks and recover drugs in 170 cases.

The preceding analysis leads to the following conclusion: In general, an enforcement strategy should be directed at the group with the highest offense rate. Citywide, since blacks were more likely to be carrying contraband than whites, the rational strategy would be to direct all stops against blacks. Consequently, a true difference in offending rates *would* justify race-based policing.

Currently, such a policy would be deemed unconstitutional; however, the Supreme Court has reversed itself often enough that relying on the Constitution to restrain a particular policy is risky at best. Again, relying on abstractions like justice and fairness risks a sea change in how society views these things. Is there some quantifiable objection to such a policy?

Table 8.   Offending Rates
(Hypothetical City, Broken Down by District)

	Black	White	Total
District A			
Carries drugs	3,900	300	4,200
Does not carry drugs	15,000	900	15,900
Totals	18,900	1,200	20,100
Offending Rate	21%	25%	
District B			
Carries drugs	100	1,700	1,800
Does not carry drugs	5,000	69,100	74,100
Totals	5,100	70,800	75,900
Offending Rate	1.96%	2.40%	

As it turns out, even if we believed such a policy to be desirable, there are purely practical reasons why it may be infeasible. In most cities, the population and crime rates are not homogeneous; there are distinct differences. Suppose our hypothetical city consists of two districts: the mostly black District A, and the mostly white District B. If we dig deeper, we might find the crime rates to break down as shown in table 8.

Even though the total numbers have not changed, a rather peculiar thing has occurred. While blacks carry drugs at a higher rate citywide, whites carry drugs at a higher rate in *both* districts! This is an example of *Simpson's paradox*, named after Edward H. Simpson, who described it in 1951. Simpson's paradox illustrates a key point: observations on aggregated data do not always translate into conclusions on disaggregated data, and vice versa. The strategy of targeting the population with the highest rate of offense is still the optimal strategy. The problem is identifying that population: the citywide data suggest we target blacks citywide; the district-by-district data suggest we target whites in District A; and it is conceivable that a precinct-by-precinct analysis could cause further revisions of where to direct our efforts.

Facially Neutral, Racially Biased?

One might argue that Simpson's paradox is only a problem if we have insufficiently detailed statistics. Thus if we *only* had the citywide data, we might reach the erroneous conclusion that targeting blacks is the more effective law enforcement strategy; with additional data, we would reach the correct conclusion that targeting whites in District A is more efficient.

Suppose we could find absolute and definitive evidence that one subpopulation broke the law more frequently than another. Would our mathematical approach lead us to conclude that race-based policing was an optimal approach to law enforcement?

Consider the problem from the officers' point of view. Their decision of whether to stop and search a vehicle or individual should be made on the basis of an evaluation of the probability that a person is committing a crime or is about to commit one; thus, an officer should only stop individuals whose characteristics and behaviors pass a certain threshold of suspicion. For example, suppose an officer has a threshold of suspicion of 25%: this means he or she will only stop a vehicle if there is at least a 25% chance that it is carrying contraband.

Of course, law enforcement officers don't have the luxury of computing the probability that a given vehicle will be carrying contraband; their judgment is based on experience and intuition. However, if we take the viewpoint that probability measures the frequency of an event in the long run, we can recover the probabilities indirectly from the fraction of stops that actually yield contraband: this is the *hit rate*. If an officer's judgment translates into only stopping vehicles that are at least 25% likely to be carrying contraband, it follows that at least 25% of the vehicles he or she stops will carry contraband.

The race of the driver might form part of the officer's assessment of the probability it is carrying contraband. If we take the viewpoint that race is just another predictive characteristic, there is nothing wrong with using race to make this assessment. What *would* be objectionable is applying different standards *because* of race. For example, if officers stopped white-driven vehicles when there was at least a 25% chance it was carrying contraband, and applied the same standard to black-driven vehicles, we should not find their actions objectionable, as the same standard is being applied to drivers of both races. On the other hand, if officers stopped white-driven vehicles when there was a 25% chance the vehicle was carrying contraband, but stopped black-driven vehicle if there was a 15% chance, this differential treatment of drivers based on their race should be cause for concern.

This is significant because, barring open admission of racism, we cannot determine whether or not an officer is using the race of the driver as part of the decision to stop a vehicle. However, it will be very easy to determine—by looking at a difference in the hit rates—whether the officer is applying different standards based on the race of the driver.

For example, in 2007, Illinois gathered data about traffic stops from more than nine hundred law enforcement agencies statewide. In that year, police throughout Illinois stopped 1,682,594 vehicles driven by whites, and 419,021 vehicles driven by blacks. They searched 10,826 (0.64%) of the vehicles driven by whites, and 8,123 (1.93%) of the vehicles driven by blacks.

Of the roughly 10,000 white-driven vehicles stopped and searched by police, about 25% contained contraband. We might conclude that whatever criteria the police used to stop white-driven vehicles, it translates into an assessment that they are 25% likely to be carrying contraband.

But at the same time, a different trend revealed itself: Of the roughly 8,000 vehicles driven by blacks that Illinois police stopped and searched, only 13% actually contained contraband. Whatever criteria the police used to stop black-driven vehicles, it translates into an assessment that they are only 13% likely to be carrying contraband.

A similar pattern was revealed by an analysis of 125,000 stops made by the New York City police between January 1998 and March 1999. The analysis indicated that 1 in 7.9 whites were arrested, compared with 1 in 8.8 Hispanics and 1 in 9.5 blacks. Again, we might interpret these as indicating that the police stopped and frisked a white person if they believed there was at least a 1 in 7.9 (roughly 12%) probability the person was engaged in a criminal activity, while they stopped and frisked Hispanics and blacks using lower threshold probabilities.

The Price of Racism

How does the difference in hit rates affect efficiency? Suppose that at any given time, every person has some probability of carrying illegal drugs. This might range from 0% (for persons who never carry drugs) to 100% (for persons who always carry drugs). As with the number of occupants per residence, these values have some probability distribution. For example, suppose a city has equal numbers of blacks and whites, but persons of different races have different chances of carrying contraband, as shown in table 9. Again, we assume a real difference in the offending rates to test the claim that a difference in offense rates justifies targeting individuals based on their race.

Table 9. Hypothetical Frequency Distribution
for Contraband

Probability of Carrying Drugs	Number of Blacks	Number of Whites
0%	400	700
10%	300	150
20%	200	100
30%	100	50

Consider two officers. The first officer is race-neutral, with a threshold of suspicion of 20%, regardless of race. In other words, this officer only conducts a stop and frisk if, in his or her assessment, there is a 20% or greater probability that the person is carrying contraband. The second officer is race-conscious, and has a threshold of suspicion of 10% for blacks and 20% for whites.

For convenience, suppose each officer is able to search everyone who meets their threshold of suspicion. The first officer, who stops and frisks only persons with a 20% or greater likelihood of carrying contraband, will eventually search 450 persons to choose from. Of the 200 blacks who are 20% likely to be carrying contraband, the officer can expect to find 40 offenders; and of the 100 blacks who are 30% likely to be carrying contraband, the officer can expect to find 30 offenders. The race-neutral officer will have searched 300 blacks and found 70 offenders, for a hit rate of about 23%.

Meanwhile, a search of the 100 whites who are 20% likely to be carrying contraband will yield 20 offenders, and a search of the 50 whites who are 30% likely to be carrying contraband will yield about 15 offenders. The race-neutral officer will have searched a total of 150 whites and found 35 offenders, and again obtain a hit rate of about 23%.

Now consider the race-conscious officer. Since this officer is searching the same group of whites as the race-neutral officer, the hit rate for whites will still be 23%. But since this officer has a lower threshold of suspicion for when to search blacks, more blacks will be searched: 600 blacks have a 10% or higher probability of carrying contraband, though only 100 will actually do so. This officer will have a hit rate of 100/600 ≈ 17%. The lower hit rate for blacks reflects the fact that this officer is using a lower threshold of suspicion: in effect, this dilutes their results and lowers their hit rate.

The result is that, since both officers can only search a limited number of persons during a shift, the race-conscious officer, with the lower hit rate, will be outperformed by the race-neutral officer. And while the preceding computation is based on a specific distribution, the result holds in general: the price of racism is reduced efficiency. If promotions, commendations, and raises are based on performance, then racism becomes its own punishment.

Stop and Frisk in New York City

In 2008, David Floyd was attempting to help a fellow tenant who had locked himself out of the apartment. Police, observing the two attempting to work the lock, approached them and frisked them. Floyd joined others in a class-action suit against the New York City police department, charging that their stop and frisk policy unfairly targeted minorities (blacks and Hispanics) at rates disproportionate to their presence in the population; Judge Scheindlin heard the case.

Between 2004 and mid-2012, New York City police stopped more than four million residents and visitors. After conducting a stop, officers are required to fill out a UF-250 form, which includes demographic information about the suspect, location, and reason for the stop. About 50% of these stops were of blacks, and 33% of Hispanics, though both groups made up about 25% of the population (each). Clearly, minorities are being stopped at a rate higher than the total of their presence in the population.

The city of New York fell back on the standard argument that minorities were being targeted at a higher rate because they made up a greater fraction of offenders. Against this argument, Jeffrey Fagan of Columbia University compiled the data from the UF-250 forms. These are summarized in table 10, which shows the distribution by race in terms of total stops, and several different hit rates: arrests; summons to appear at the police precinct or court; or illicit items found.

Table 10. Hit Rate for New York City's Stop and Frisk, 2004–Mid-2012

	Total Stops	Arrests (%)	Summons (%)	Guns (%)	Other Weapons (%)	Contraband (%)
White	435,036	6.00	6.34	0.08	1.37	2.27
Black	2,289,156	5.68	6.43	0.19	0.96	1.76
Hispanic	1,361,926	5.98	6.61	0.11	1.12	1.71
Other Race	324,102	4.95	6.02	0.08	0.77	1.37

Note that whites have substantially higher hit rates for contraband and other weapons. It's not completely clear how to interpret this fact since the reason an officer decides to conduct a stop and frisk might not coincide with the items recovered: an officer might suspect a person is carrying illegal drugs, but discover an illegal weapon instead. However, it does suggest that officers have a lower threshold of suspicion for blacks and Hispanics. Fagan's data also showed that between 2004 and 2009, blacks were about 8% less likely to be subject to further enforcement action than whites, *and* this disparity was larger the greater the black population of the precinct, further supporting the argument that police had a lower threshold of suspicion for blacks than for whites. As a result, Judge Scheindlin ruled against the city. On January 30, 2014, Mayor Bill de Blasio announced that the city would not appeal the decision.

While Judge Scheindlin denied the relevance of the effectiveness of the stop-and-frisk policy, her decision was in part based on an inevitable consequence of race-based inefficiency: a differential hit rate. It follows that those who practice a race-based policy will suffer in two ways: first, from a lowered overall efficiency, and second, from a differential hit rate that makes it easy to identify when a race-based policy is in operation. Racism is simply not efficient: those who practice it do so to their detriment.

Reverend Thomas Bayes and the Law

> No warrants shall issue, but upon probable cause, supported by Oath or
> affirmation, and particularly describing the place to be searched, and
> the persons or things to be seized.
>
> —U.S. CONSTITUTION, AMENDMENT IV

On September 11, 2001, the world changed in a dramatic fashion when 19 terrorists hijacked four planes and succeeded in crashing them into three inhab-
ited buildings: the two World Trade Center towers in New York City, and the
Pentagon in Alexandria, Virginia. The days and years to come made clear
what farsighted analysts had predicted as far back as the early 1970s: the most
serious threats to national security came not from other nation-states, but
from groups motivated by ideology.

In the wake of the terrorist attacks of September 11, 2001, the Defense
Advanced Research Projects Agency (DARPA) initiated a project known as
Total Information Awareness (TIA, later renamed Terrorism Information
Awareness). TIA had two components. One was to try to construct a profile
of potential terrorists. The second was to use available information to identify potential terrorists in the general population *before* they could commit
an act of terrorism.

How could this be done? The ever-growing computerization of our society gives us an unprecedented amount of information about a person's
day-to-day activities. The emerging field of *data mining* applies statistical
techniques to this data to recover patterns that might not be observable
otherwise.

The basic problem of data mining can be expressed as follows. Suppose we
have a large amount of information about a group of people that we wish to

classify into distinct categories. We might have information of all persons who do online shopping, and seek to separate them into persons likely to buy this book, and those unlikely to buy this book. Or, in the case of TIA, we want to split the entire population into those likely to be terrorists and those unlikely to be terrorists. Our goal is to write a series of questions whose answers will allow us to separate the group into two or more sets, one of which has a higher-than-average presence of the persons of interest.

To focus on a single aspect of a person's life, consider their cell phone usage. We might compare the cell phone usage patterns of known terrorists to the cell phone usage patterns of known non-terrorists, and we might find that 90% of the terrorists have average call lengths of less than two minutes. Given unlimited information, we could identify similar patterns. For example, we might discover that 90% of terrorists (not necessarily the same 90%) receive calls from an average of three or fewer people each month, and that 90% of them (again, quite possibly a set different from the first two) make calls mostly in the early morning or late afternoon.

Some of these characteristics fit our intuitive notion of how terrorists communicate: for example, we would not be surprised to find that terrorists make short cell phone calls to a limited number of people. But a sufficiently detailed search of call patterns will doubtless reveal some unexpected patterns. For example, we might discover that 90% of terrorists order take-out food at least twice a month. There is no obvious connection between being a terrorist and a predilection for take-out food—but research is most valuable when it discovers things that we *didn't* expect to find.

In order to be useful, this information must come from positively identified terrorists. We might analyze the prior activities of those who have been convicted of the crime of terrorism. As convicted felons, this group would have reduced expectations of privacy, so obtaining information about their prior activities raises few constitutional issues.

Implementing TIA requires virtually unlimited access to everyone's personal information to identify potential terrorists. DARPA's failure to follow certain protocols for projects of this nature led to widespread and vehement public opposition to "Big Brother" style monitoring; DARPA shut down the program in 2003. Despite this, in 2006 the Department of Homeland Security reported that some of its units were engaged in, or were preparing to engage in, data mining to identify potential terrorists, and in 2007, the National

Security Agency launched PRISM to collect telecommunication information on an unprecedented scale.[1]

There are two objections to data mining as a method of finding potential terrorists. The first involves privacy issues: just how much personal information can the authorities obtain about a person without a warrant? The issue is blurred because not all personal information is well-protected by the 4th Amendment. For example, if you buy a pair of shoes at a store, the fact that everyone in the store (and especially the sales clerk) knows what pair of shoes you are buying effectively negates the requirement for a warrant to obtain information about the type of shoes you buy.

The general principle is that information conveyed to a third party is not generally covered under the 4th Amendment's prohibition of unreasonable search. Consider a cell phone conversation. It is impossible to conceal the number we are calling or the length of the call from the phone company: thus we have no expectation of privacy regarding this type of information (known as *metadata*).

We do have a reasonable expectation of privacy for the *content* of the communication. We might ask why the content of the phone call is different. After all, when we speak into a phone receiver, the sounds we make must be conveyed to the phone company for transmission to the recipient. There are various legal theories for why content is protected, but we can take a purely pragmatic point of view: if the information *can* be concealed from a third party, then we have a reasonable expectation of privacy.

It is impossible to conceal the phone number you are calling from the phone company: if you do so, it won't be able to complete the call! On the other hand, the content can be concealed by talking in code: an innocuous phrase like "Don't forget to pick up cream" could be the signal to launch a terrorist attack—or to meet an unfaithful spouse at the usual location. Since we can effectively conceal the content, we have a reasonable expectation of privacy.

The second concern is that TIA envisioned being able to detect a terrorist plot in process by focusing on a pattern of activities. But most of the activities in question would be innocuous. For example, some of the 9/11 hijackers purchased first-class airline tickets. However, buying first-class airline tickets is

---

1. Edward Snowden's revelation, in 2013, of the scale of PRISM raised some as-yet-unresolved constitutional issues over the permissible extent of such surveillance.

not a crime, so we might wonder if it is even possible to use perfectly legal activities as grounds for a search or arrest warrant. The answer is that it depends on the activities and characteristics. For example, in 1984, Drug Enforcement Administration (DEA) agents noted that airline passenger Andrew Sokolow had paid $2,100 cash for his airline tickets; had returned to Honolulu after a very brief stay in Miami; had checked no bags; and appeared nervous. These characteristics matched a profile of drug smugglers ("mules") constructed by the DEA, so they stopped Sokolow in the Honolulu airport and searched his bags, finding a kilogram of cocaine.

Sokolow argued that the basis for the search was unreasonable: there is nothing illegal about buying airline tickets with cash, or staying in Miami for a short period of time, or being nervous before a flight, or not checking bags. The Supreme Court heard Sokolow's case in 1989, and in a 7 to 2 decision, ruled that a collection of innocuous characteristics can, when observed in a single individual, be legitimate grounds for further investigation of that individual.

Before we proceed, we note that in principle there is no difference between "paying cash for an airplane ticket" and "ordering take-out food." Both of these are legal activities, and if the one can be used as a basis for probable cause, there is no a priori reason why the other should not be.

Priors and Posteriors

To evaluate the effectiveness of our approach, let's assume that we have obtained information like "90% of terrorists have average call lengths of less than two minutes." The obvious use of this information is to focus on persons who share similar characteristics, thus, persons whose average call lengths are less than two minutes would "fit the profile" and be subject to further investigation.

The crucial question is: can this information help us identify potential terrorists? In effect, we are presented with a person who might (or might not) be a terrorist. We measure this uncertainty using probability. For example, suppose that in the general population, one person in 10,000 is a terrorist, in the sense that he or she is willing to commit (not just advocate) acts of violence for political reasons. Based on this information and nothing else, we can say that the *prior probability* that a person is a terrorist is 1/10,000.

This probability will almost certainly change if we know a little more about the person. For example, if 90% of terrorists have average cell phone call lengths that last less than two minutes, then knowing that a person has an average cell phone call length of less than two minutes should allow us to form a revised estimate of the probability that he or she is a terrorist.

This revised estimate is known as the *posterior probability*. Posterior probabilities are also known as *conditional probabilities*, and break down into two components: the fact we know with certainty, and the fact we are uncertain about.

For example, consider the information that 90% of terrorists have average call lengths of less than two minutes. In our scenario, this fact is obtained by examining the call patterns of people known to be terrorists; thus, "a person is a terrorist" is the fact known with certainty. What we are uncertain about is whether or not their average call length is less than two minutes. The information given corresponds to the probability that *a person has an average call length of less than two minutes*, given that we know with certainty that *the person is a terrorist*.

On the other hand, suppose we have a list of people whose average call length is less than two minutes, and we are interested in knowing whether any of them are terrorists. It follows that we want to know the probability that *a person is a terrorist*, given that we know with certainty that *the person's average call length is less than two minutes*.

Note that the position of the italicized clauses has switched. We know the first probability (it is 90%), but we want the second probability. The *fallacy of the transposed conditional* occurs when we assume that the two probabilities are the equal or nearly so. Rather distressingly, the fallacy of the transposed conditional appears in criminal proceedings with such regularity that another name for it is the *prosecutor's fallacy*. The most notorious instance of this (but, by no means, the first, last, or only) involved a 1964 robbery of an elderly California woman. Eyewitness accounts described the assailant as a blond woman with a ponytail, who was observed to enter a yellow car driven by a black man with a beard and mustache. The police eventually charged Malcolm and Janet Collins with the crime.

At Malcolm's trial, the prosecutor noted that Malcolm and Janet Collins were an interracial couple matching the characteristics of the perpetrators, as given by the eyewitnesses. The prosecutor gave an example (discussed below) that suggested the probability of there being another interracial couple consisting of a blond woman with a ponytail, a black man with a beard and mustache, and driving a yellow car, was 1 in 12,000,000. Thus, the prosecutor argued, the probability that the Collinses were innocent was remote.

The California State Supreme Court heard *People v. Collins* (1968) and overturned the conviction on several grounds, not the least of which was the probability argument itself was flawed; we will discuss the flaws below. More important, the California Supreme Court recognized that even if the probabilities were correct, the probability was irrelevant: the 1/12,000,000 probability obtained by the prosecutor is in fact the probability that *a person matches the characteristics of the perpetrator*, given that we know with certainty that *the person is innocent*. But the relevant information is the probability that *a person is innocent*, given that we know with certainty that *the person matches characteristics of the perpetrator*.

Collins has been interpreted by lawyers and judges alike as rejecting explicit probability computations as evidence of guilt. For example, if the DNA found at a crime scene matches the DNA of the defendant, a prosecution witness might cite the *random match probability*: the probability that the DNA taken from an innocent person matches the DNA found at a crime scene. Generally speaking, this value is exceedingly low: for example, 1 in 100 trillion. This corresponds to the probability that *a person matches the DNA found at the crime scene*, given that we know with certainty *the person is innocent*. In other words, it is extremely unlikely. However, the prosecution and, by extension, the witness for the prosecution, must refrain from stating or implying that this is the probability *the person is innocent*, given that *the person matches the DNA found at the crime scene*.

In 1994, for example, a Nevada jury convicted Troy Brown of child rape partly on the basis of a match between his DNA and DNA found at the scene of the crime. Renee Romero, expert witness for the state (Nevada), testified that the probability that another person matched the DNA was one in three million. However, the prosecutor appeared to equate this with a one in three million chance that the DNA did not come from Troy Brown, which is a variant of the prosecutor's fallacy. This led the Court of Appeals for the Ninth Circuit to grant Brown a writ of habeas corpus, the first step in overturning the conviction.

The state appealed, and in *McDaniel v. Brown* (2010), the Supreme Court set an important standard. While it objected to some of the probability testimony introduced at Brown's trial, it noted that other evidence would have been sufficient for conviction. Consequently, it upheld the original conviction.

Positives and Negatives, True and False

Understanding and identifying the prosecutor's fallacy is particularly important when considering a profile to identify a criminal or terrorist. For example, suppose we have a profile that can identify a terrorist 99% of the time. If a person matches the profile, what should we do? In the case of drug smugglers, it's easy enough to pull the person aside and search his or her belongings. Of particular significance in *Sokolow* and similar cases was the fact that such a search was minimally invasive and inconveniencing: a person wrongfully accused could clear themselves in a few minutes.

In contrast, a person identified as a terrorist has a much more difficult task: if the authorities find no evidence of terrorism, they might conclude they are dealing with an innocent person—but they could just as easily conclude they are dealing with a very clever terrorist. Unlike the situation in *Sokolow*, there is almost no limit to how invasive and inconvenient an investigation can become.

How do we balance civil liberties with security? In the absence of an objective measure, this comes down to trading abstractions. All too often, it comes down to a trade of one group's civil rights for another group's security—with the latter group eager to make the trade, and the former group just as eager to refuse it. A quantitative analysis offers a much more useful way to assess these tradeoffs.

It's useful to view terrorism and other criminal activities as diseases: treating the effects and punishing the perpetrators is desirable, but an even better use of resources is to prevent terrorism and crime in the first place. A medical test will reveal the presence of a condition so that treatment can begin as soon as possible; a profile will (potentially) reveal the inclination of a person toward terrorism or crime. In the parlance of the medical field, a test returns a *positive result* if it indicates the presence of a condition, and a *negative result* otherwise. However, biology is complicated, and there are many reasons why a test result might not reflect the true state of the world. A *true positive* occurs when a condition is present *and* the test returns a positive result, while a *true negative* occurs when the condition is absent *and* the test returns a negative result.

Of concern are *false positive* and *false negative* results. A false positive occurs when a test indicates the presence of a condition (the positive result), but the condition is not present (the false result). One of the better studied examples of this involves tests for opiate drug use: a poppyseed bagel contains enough morphine to cause a person to test positive for heroin use. On a number of occasions, otherwise law-abiding citizens were accused of drug use, and were only cleared after extensive investigations accompanied by expensive litigation. A false negative can also occur, when the test indicates the condition is absent when it is actually present. A false negative could have potentially catastrophic results: a false negative for skin cancer, which is easily treatable in its early stages, could allow it to progress to the point where it is difficult or impossible to treat.

The accuracy of a test is measured by two values. First is the *sensitivity*, which is the fraction of persons with the condition for whom the test will indicate the presence of the condition. Next is the *specificity*, which is the fraction of persons *without* the condition whom the test indicates lack the condition. In our terrorism example, the sensitivity corresponds to the probability that a person, known to be a terrorist, fits a profile. Mathematicians, like most normal people, are averse to complex grammatical constructions like "the probability that a person, known to be a terrorist, fits a profile." Thus they introduce notation to represent complex statements. The notation $P(A \mid B)$, sometimes read "the probability of A *given* B," can then be interpreted as the probability that A is true, given that we know that B is true.

We'll let $T =$ a person is a terrorist, and $L =$ a person fits a profile for terrorists. Thus $P(L \mid T)$ is the probability that a person fits the profile, given that we know the person is a terrorist. This corresponds to the sensitivity of the test. Meanwhile, $P(\text{not } L \mid \text{not } T)$ is the probability that a person does not fit the profile, given that we know the person is not a terrorist. This corresponds to the specificity of the test. To determine these values, we must begin with a sample of known terrorists and a sample of known non-terrorists. As noted earlier, we can determine $P(L \mid T)$ easily, by researching the activities of known terrorists.

$P(\text{not } L \mid \text{not } T)$ is more difficult: we must obtain a sample of non-terrorists. However, for the value we obtain to be meaningful, our non-terrorists must be representative of the population we intend to apply the test to. Thus, if we wish to use our profile on airline passengers in order to construct a "no-fly" list of potential terrorists, we must find a group of people representative of airline passengers.

Suppose our test is 99% sensitive, so that the probability that a person fits the profile, given that the person is in fact a terrorist, is 99%: $P(L \mid T) = 0.99$. We also know that the probability that a person does *not* fit the profile, even though we know the person is in fact a terrorist, is 1%: $P(\text{not } L \mid T) = 0.01$.

Next, suppose the test is 99% specific, so that e the probability that a person does not fit the profile, given that the person is not in fact a terrorist, is 99%: $P(\text{not } L \mid \text{not } T) = 0.99$. And as before, this means that the probability that a person fits the profile, even though the person is not a terrorist, is 1%: thus $P(L \mid \text{not } T) = 0.01$.

What we are interested in is the probability that a person who fits the profile is a terrorist: in other words, we know a person fits the profile, but we are unsure of whether or not they are a terrorist. This corresponds to $P(T \mid L)$. While we know $P(L \mid T)$, we also know that we *cannot* assume any relationship between $P(L \mid T)$ and $P(T \mid L)$: doing so leads to the prosecutor's fallacy.

How can we find $P(T \mid L)$? Let us continue to suppose that 1 in 10,000 of the general population is a terrorist, and suppose we apply our profile to a group of ten thousand people that includes 1 terrorist and 9,999 non-terrorists.

The single terrorist has a 99% chance of matching the profile. Thus the profile is almost certain to identify the lone terrorist. However, remember we have applied our profile to all 10,000 persons, and 1% of non-terrorists also fit the profile; thus in our group, there will be about 100 non-terrorists who fit the profile. Altogether, we have $100 + 1 = 101$ persons who fit the profile, only 1 of whom is a terrorist. This gives us a less than 1% chance that a person matching the profile is in fact a terrorist.

From a civil liberties point of view, any misidentification is problematic, since it would threaten the civil rights of those wrongfully accused. Against this we have the argument that terrorism is such a threat that violating the civil rights of many people to prevent a terrorist act is a worthy tradeoff. Again, this type of argument often comes down to trading one group's civil liberties for another group's security.

The quantitative approach offers a much more useful argument against the profile: if we use it, we must investigate 100 non-terrorists for every terrorist we catch. But investigating an innocent person takes time, personnel, and money. In cases like *Sokolow*, where the investigation takes a few minutes at most, this waste is nominal; but if we must do a more in-depth investigation for *every* suspect, the wasted effort can overwhelm law enforcement resources.

To avoid this waste, we must focus on a group more likely to contain terrorists. For example, suppose we focus on a group where 1 in 10 of the members is a terrorist. In this case, if we tested 1,000 people, 100 would be terrorists, and the profile would identify 99 of them. Of the 900 non-terrorists, 1% (9) would also match the profile. Thus $99 + 9 = 108$ would match the profile.

But in this case, fully 92% of those would be terrorists. Consequently, an equivalent investment of law enforcement resources (investigating about 100 persons) has a far greater return.

Another case in which this analysis would be useful is drug testing for recipients of public assistance. Since the 1980s, stories have circulated about welfare queens dining on caviar in their penthouse apartments, leading to claims that many on public assistance are receiving benefits fraudulently; demands for increased scrutiny of those applying for assistance; and monitoring of those receiving benefits.

In May 2011, Florida passed a law requiring persons applying to the TANF (Temporary Assistance for Needy Families) program to take and pass a drug test before applying for benefits; other states have considered, and some have passed, similar legislation. Within a few months, the program was halted temporarily pending a review of the constitutional issues, and on December 31, 2013, U.S. District Judge Mary Scriven halted the program permanently, calling it a warrantless, suspicionless search.

Scriven's decision was based on abstract constitutional issues. However, Bayes's theorem offers a more quantitative approach. The specificity and sensitivity of drug tests vary considerably, but a number of studies suggest that tests for marijuana usage (the test at issue in Florida and other states) have specificities and sensitivities in the 70–90% range. Suppose both are 95%. Roughly 10% of the general population reports occasional usage of marijuana.

As in our terrorism example, let us suppose we test 10,000 persons for marijuana usage. About 1,000 would in fact be users, and 95% of them—950— would be identified as such. Meanwhile, 9,000 would not be marijuana users, but 5% of them—450—would be misidentified as such. A total of 1,400 persons would be identified as marijuana users, but nearly one-third of those persons would be innocent.

From a civil liberties point of view, there is another problem. A test for a medical condition, like a skin test for melanoma, can always be repeated to weed out false positives. False positives are common enough in medical testing that few professionals are willing to convey bad news to patients: if a test returns a positive result, the invariable response is to call the patient in to "run some more tests."

But suppose a test identifies a person as a drug user. A later test will, at best, show that the person has not used drugs between the first and second tests; it will not clear them of the accusation of having used drugs at the time of the first test. Persons trying to clear themselves of wrongful accusations of drug use are in the position of trying to prove a negative: a single false positive result can produce lasting and irreparable damage to a person's life.

Let us disregard the constitutional and practical issues associated with false accusations of drug use, and consider another factor. Florida's law was in operation for four months before the initial injunction was granted. During that time, Florida tested 4,086 applicants, and found 108 marijuana users. Even if we assume that every single one of these was a true positive, it translates into a usage rate of 2.6%. The low yield suggests that this approach fails the "individualized suspicion" requirement of the 4th Amendment: if this test was being used to identify drug users, it would surely be ruled unconstitutional on these grounds. More pragmatically, it fails the rational requirement that the probability a search will find evidence of wrongdoing should be sufficiently high to justify the time, effort, and above all money put into the search.

In Florida, the persons confirmed by the test as non-users had their test expenses reimbursed by the state, so for a price of $118,140, the state of Florida kept 108 persons off TANF. Since a person can receive benefits for up to five years, this appears to be a worthy use of money.

However, the real question is whether $118,140, roughly $30,000 each month, could have kept *more* than 108 persons off TANF. For example, suppose it is true that many of those receiving TANF are doing so fraudulently. Then $30,000 per month might be better spent staffing a fraud unit, with perhaps a half-dozen investigators, to detect such individuals. If each investigator found just one case of fraud a week, they would have outperformed the drug testing program.

Bayes's Theorem

Suppose for a moment that a large fraction of TANF recipients were drug users. This might have altered the court's decision; as in *Sokolow,* one could argue that an innocuous factor (applying for TANF benefits) could be the basis for a *reasonable* search (mandatory drug test).

The main problem with this argument is that it requires that we know the actual yield of the search. Thus, in the TANF case, the yield was so small (2.6%) that the court ruled that the search was unreasonable. However, the yield could not have been known beforehand, so 4,086 TANF applicants were tested before the preliminary injunction was granted: in some sense, the constitutional rights of the 4,086 persons were violated.[2] Is there some way of determining the yield beforehand? Unfortunately, criminals rarely identify themselves on census surveys or tax forms, so we cannot determine the yield of a test in advance. However, we can gain some insight as follows.

In general, we have two related probabilities: the probability that *A occurs*, given that we know with certainty that *B has occurred*, which we write as $P(A \mid B)$; and the probability that *B occurs*, given that we know with certainty that *A has occurred*, which we write as $P(B \mid A)$. The prosecutor's fallacy is based on assuming a simple relationship between them (equality, in most cases, though assuming they are of a similar magnitude is also common).

The first statement of the relationship between the two probabilities was given by Thomas Bayes (1701–1761), a Presbyterian minister who also studied mathematics. In *An Essay Towards Solving a Problem in the Doctrine of Chances,* read to the Royal Society in 1763 (two years after his death), Bayes gave a version of what is now known as *Bayes's theorem*:

$$P(A \mid B) \times P(B) = P(B \mid A) \times P(A)$$

where $P(A)$ and $P(B)$ correspond to the probability that A has occurred or that B has occurred.

In our case, we might take $T =$ a person is a terrorist, and $C =$ a person has an average call length of less than two minutes. Then our assumptions give us $P(C \mid T) = 90\%$. We also assumed $P(T) = 1/10{,}000$. To find $P(T \mid C)$, the probability that a person is a terrorist *given that we know with certainty* that their average call length is less than two minutes, we need to find $P(C)$, the probability that a person has an average call length of less than two minutes. For illustrative purposes, suppose that $P(C) = 20\%$: in other words, 20% of cell phone users have average call lengths of less than two minutes.

Bayes's theorem gives us

$$P(C \mid T) \times P(T) = P(T \mid C) \times P(C).$$

Substituting in our values gives us:

$$(0.90) \times (0.0001) = P(T \mid C) \times (0.20).$$

---

2. This *includes* the 108 persons actually revealed to be drug users. Just because a search is successful does not make it constitutional.

Thus we find $P(T|C) = 0.00045$. In other words, the probability that a person is a terrorist, given that we know with certainty that their average call length is less than two minutes, is 0.00045, or about 1 in 2200. This supports our intuition that persons with average call lengths of less than two minutes are not likely to be terrorists.

Of course, this value relies on $P(T) = 0.0001$ and $P(C) = 0.2$. The latter, which corresponds to the fraction of cell phone users that have average call lengths of less than two minutes, can be found easily from phone company records (and such investigations need not even raise privacy issues, since neither the identity of the caller nor the receiver need be known). $P(T)$, on the other hand, is very likely unknown for the population in question.

To avoid the problem of unknown probabilities, we can use an alternative measure of probability: the odds of an event. The odds are the ratio of the frequency with which an event occurs to the frequency that it does not occur. For example, if Horse A wins 3 out of 10 races, and loses 7 out of 10 races, the odds that it wins are 3/10 to 7/10, or 3 to 7. In common language, this fact can be stated in a number of different ways:

- The odds of the horse winning the race are 3 to 7.
- The odds of the horse winning the race are 3 to 7 *for*.
- The odds of the horse winning the race are 7 to 3 *against*.

The last phrasing is particularly common with low-frequency events. Consequently (based on our cell phone example), the odds that someone is a terrorist, given that their average call length is less than two minutes, should be given as 1 to 2200, since we find 1 terrorist among 2200 non-terrorists. But instead, we might say the odds are 2200 to 1 *against*.

Now consider again our cell phone usage example, but this time assume that we only know the probability that a terrorist has an average call length of less than two minutes (90%), and the probability that a non-terrorist has an average call length of less than two minutes (20%). Both of these values may be regarded as being easily obtainable: the first from research on identified terrorists, and the second from an analysis of a collection of cell phone records of representative non-terrorists.[3]

---

3. If the number of terrorists in our population is small compared to the population, this number is not substantially different from the fraction of *all* cell phone users whose average call

Bayes's theorem gives us

$$P(T\,|\,C)\times P(C)=P(C\,|\,T)\times P(T)$$

where we again use C=a person has an average call length of less than two minutes, and T=a person is a terrorist. Bayes's theorem *also* gives us

$$P\,(\text{not }T\,|\,C)\times P(C)=P(C\,|\,\text{not }T)\times P(\text{not }T).$$

Dividing the two equations by each other gives us

$$\frac{P(T\,|\,C)}{P(\text{not }T\,|\,C)}=\frac{P(C\,|\,T)}{P(C\,|\,\text{not }T)}\times\frac{P(T)}{P(\text{not }T)}$$

This equation can be broken into three parts.

First, $\dfrac{P(T)}{P(\text{not }T)}$ is the ratio of the probability that a person is a terrorist to the probability that a person is not a terrorist: in other words, these are the odds that a person is a terrorist. More specifically, they are the odds that a person is a terrorist, given that we know nothing else about them. Since these odds are based on the prior probabilities, they are known as the *prior odds*.

Next, $\dfrac{P(T\,|\,C)}{P(\text{not }T\,|\,C)}$ is the ratio of the probability that a person is a terrorist, given that we know with certainty their average call length is less than two minutes, to the probability that a person is not a terrorist, given that we know with certainty that their average call length is less than two minutes. These are the odds based on the posterior probabilities, so they are known as the *posterior odds*. Like the posterior probabilities, we can view this value as a revised estimate, based on additional information.

Prior and posterior odds are related by the third term $\dfrac{P(C\,|\,T)}{P(C\,|\,\text{not }T)}$, called the *odds multiplier*. Note that neither the prior nor the posterior odds can be found without knowing the prevalence of terrorists in the population. However, the odds multiplier itself can be found using easily available data. In our cell phone example, we assumed that the probability that a person's average cell phone calls are less than two minutes, given that the person is a terrorist, was 90%: thus $P(C\,|\,T)=0.90$. We also have the probability that a person's

---

length is less than two minutes. Provided that only the compiled data are used, no privacy issues arise from obtaining this information.

average cell phone calls are less than two minutes, given that the person is not a terrorist, was 20%: thus $P(C|\text{not } T) = 0.20$. This means that our odds multiplier is $\dfrac{P(C|T)}{P(C|\text{not } T)} = \dfrac{0.90}{0.20}$, or 4.5, and we have:

$$\frac{P(T|C)}{P(T|\text{not } C)} = 4.5 \times \frac{P(T)}{P(\text{not } T)}$$

Whatever the odds that a given person is a terrorist, the fact that their average cell phone calls are less than two minutes in length increases those odds by a factor of 4.5.

Non-Exclusionary Non-Matches

The importance of this step is that the odds multipliers work in both directions: if the odds are increased by matching a characteristic, they are decreased by a non-match. In theory, non-matches should be particularly important in criminal investigations. However, practice does not match theory.

For example, Brandon Mayfield, a lawyer in Portland, Oregon, was detained by the FBI for 19 days following the March 11, 2004 terrorist bombings of commuter trains in Madrid, Spain. Mayfield's arrest was based on a partial fingerprint recovered by Spanish authorities on a bag of detonating devices. On the one hand, Mayfield was one of at least twenty people in the United States whose fingerprints were a partial match to those recovered by Spanish authorities. But the narrative of Mayfield's life included other factors that distinguished him from the other nineteen: He was an army veteran; he had married an Egyptian woman and converted to Islam; and provided legal services in a child custody case for a man sentenced to prison for trying to travel to Afghanistan to join the Taliban.

Even if we grant that Mayfield matched some factors in a terrorist profile, the non-matches are even more profound: Mayfield had not left the country since 1994; he had no explosives expertise; he had no access to bomb-making equipment; he had no known contacts with other terror suspects. Working without a quantitative approach, a law enforcement agent would balance the matches (fingerprints and life narrative) against the non-matches (no expertise or access to materials, no contacts with other terrorists, no international travel). However, a number of studies show that people tend to underestimate the significance of non-matches, even to the point of ignoring them entirely. Bayes's theorem offers us a way to quantify the significance of non-matches.

In our continuing example, we find that having short cell phone conversations will multiply the odds that a person is a terrorist by 4.5. It's tempting to believe that since having short cell phone conversations multiplies the odds by 4.5, then not having short cell phone conversations divides the odds by the same factor. However, the relationship between the multipliers is rarely this direct. Instead, we have:

$$\frac{P(T \mid \text{not } C)}{P(\text{not } T \mid \text{not } C)} = \frac{P(\text{not } C \mid T)}{P(\text{not } C \mid \text{not } T)} \times \frac{P(T)}{P(\text{not } T)}$$

The odds multiplier for a non-match will be $\dfrac{P(\text{not } C \mid T)}{P(\text{not } C \mid \text{not } T)}$. By our assumptions, 10% of terrorists have average call lengths of two minutes or greater, as do 80% of the general population; it follows that $\dfrac{P(\text{not } C \mid T)}{P(\text{not } C \mid \text{not } T)} = \dfrac{0.10}{0.80} = 0.125$. It follows that if we suspect a person is a terrorist, but find that their average call lengths last two minutes or more, then the odds the person is a terrorist have decreased by a factor of 0.125 (one-eighth).

Multiplying the Multipliers

The key principle of the *Sokolow* ruling is that a *collection* of innocuous factors could form the grounds for probable cause. Thus we could not use *just* the fact that a person's cell phone calls lasted an average of less than two minutes as evidence that a person is a potential terrorist, any more than the DEA would use *just* the fact that a person paid cash as evidence he or she was a drug mule.

To proceed, it is helpful to introduce the concept of *independence*. A conditional probability can be viewed as a revised assessment of the probability of an event, *given* that we know another one has occurred. For example, if we are interested in knowing the probability that a snowstorm will delay a plane flight, then our evaluation of this probability will change significantly if we know that the plane is leaving Miami, Florida. Because our probability estimate changes given the additional information, we say the events are *dependent*. Independence and dependence work both ways. Knowing a plane is leaving Miami alters our evaluation of the probability that it is delayed by a snowstorm. It is also true that knowing a plane was delayed by a snowstorm will alter our assessment of the probability that it left Miami.

In some cases, two events give us no information about each other. For example, our assessment of the likelihood that a snowstorm delays a plane flight will probably not change if we are told that the flight is on a Tuesday. In this case, we say that the events are *independent*. The *multiplication principle* states that the probability that a set of independent events occurs at the same time is the product of the probability of each event occurring.

It is critical to the application of the multiplication principle that the events actually be independent. In *Collins*, the prosecutor's probability was obtained as follows: assume for illustrative purposes that the probability that a black man has a beard is 1/10; the probability that a man has a mustache is 1/4; the probability that a woman has a ponytail is 1/10; the probability that a woman is blond is 1/3; the probability that a car is yellow is 1/10; and the probability of an interracial couple was 1/1000 (this was 1964), then the probability that an innocent couple matched all characteristics was

$$\frac{1}{10} \times \frac{1}{4} \times \frac{1}{10} \times \frac{1}{3} \times \frac{1}{10} \times \frac{1}{1000} = \frac{1}{12,000,000}.$$

Multiplying probabilities only works when the events are independent. But it seems that the probabilities of some of the events will be revised when we know that others have occurred: the most obvious is that whatever we believe to be the probability that a black man has a mustache, we would surely change our estimate of the probability if we knew that he had a beard.

The requirement that events be independent in order for the multiplication of their probabilities to be relevant complicates the mathematical problem of constructing and evaluating a profile. There are two approaches. The more precise is to use the correct conditional probabilities. The problem is that in a data mining application, there might be hundreds of characteristics, and finding the correct conditional probabilities means finding groups of people that meet any particular set of characteristics—and there may be some sets of characteristics that are met by nobody. This does not mean they will *never* occur: it just means that our sample isn't large enough. Thus, a string of five consecutive "heads" results could occur by flipping a coin, but if you only flipped a coin 20 times, you might not see such an occurrence. Thus in practice, obtaining the correct conditional probabilities might not be feasible.

A second approach is to treat the characteristics as actually being independent. This is risky, but we can use it to develop a provisional profile, making a good faith effort to avoid obviously dependent characteristics. For example, if we use "makes short cell phone calls" as one of our characteristics, we should probably not use "has a pay-per-minute cell phone plan" as another, as the two would seem to be dependent. We can use *either* of two dependent characteristics, so long as we avoid using *both*.

Suppose we find that only 1% of non-terrorists receive calls from an average of three or fewer people each month, while 90% of terrorists do so. Thus letting $F =$ a person who receives calls from an average of three or fewer people each month, we have $P(F \mid T) = 0.9$ and $P(F \mid \text{not } T) = 0.01$. This gives an odds multiplier of $\dfrac{P(F \mid T)}{P(F \mid \text{not } T)} = 90$. Since we must also consider the possibility a person of interest does not receive calls from an average of three or fewer people each month, we also need $\dfrac{P(\text{not } F \mid T)}{P(\text{not } F \mid \text{not } T)}$. In this case, we have $P(\text{not } F \mid T) = 0.1$ and $P(\text{not } F \mid \text{not } T) = 0.99$, so $\dfrac{P(\text{not } F \mid T)}{P(\text{not } F \mid \text{not } T)} \approx 0.101$.

Finally, suppose we find that 40% of non-terrorists order take-out food at least twice a month, while 90% of terrorists do so. Letting $M =$ a person who orders take-out food at least twice a month gives us $P(M \mid T) = 0.9$, $P(M \mid \text{not } T) = 0.4$, $P(\text{not } M \mid T) = 0.1$, $P(\text{not } M \mid \text{not } T) = 0.6$, and $\dfrac{P(M \mid T)}{P(M \mid \text{not } T)} = 2.25$, $\dfrac{P(\text{not } M \mid T)}{P(\text{not } M \mid \text{not } T)} \approx 0.167$. We summarize these odds multipliers in table 11. Now consider three persons X, Y, and Z. Person X meets all three characteristics. Person Y meets C and F but not M, and Z meets F and M but not C. We must also consider all evidence: both the evidence that tends to implicate (the matches) and the evidence that tends to exonerate (the non-matches). This

Table 11.   Odds Multipliers for Profile Characteristics

	Match	Don't Match
Average call length less than 2 minutes (C)	4.5	0.125
Receive calls from three or fewer people each month (F)	90	0.101
Order take-out food at least twice a month (M)	2.25	0.167

has nothing to do with fairness—it has everything to do with efficiency, since ignoring exonerating evidence causes us to waste time and resources on people who are not likely to be terrorists. Once again, focusing on efficiency, which can be measured objectively, forces us to guard civil liberties and avoid unjust accusations.

For X, who meets all characteristics, the odds multiplier will be $4.5 \times 90 \times 2.25 \approx 900$. For Y, who meets C and F but not M, the odds multiplier will be $4.5 \times 90 \times 0.167 \approx 70$. Finally, for Z, who meets M and F but not C, the odds multiplier will be $0.125 \times 90 \times 2.25 \approx 25$. It's tempting to set some threshold value for the odds multiplier—say a hundred—above which we would have probable cause for further investigation. Given the odds multipliers above, we might focus our resources on X and ignore Z. However, this runs afoul of the *base rate fallacy*: if the probability of being a terrorist is low, it will remain low even with a large odds multiplier.

For example, standard law enforcement techniques independent of our profile might identify, from the general population, a group of potential terrorists. We might focus on those who've set up web pages giving step-by-step instructions on how to commit acts of terrorism. In this group, the base rate (the fraction of the group that are terrorists) would be relatively high, say 50%, so the prior odds that a person is a terrorist will be 1 to 1. On the other hand, the base rate for terrorists in the general population will be much lower: for example, perhaps 1 in 10,000 of the general population is a terrorist. Then the prior odds are 1 to 10,000.

Suppose, for argument's sake, that Z came from the first group and X came from the second. Since Z's characteristics correspond to an odds multiplier of 25, the odds that Z is a terrorist go from 1 to 1, to 25 to 1. Meanwhile, the *probability* that Z is a terrorist goes from 50% to 96%: from a reasonable likelihood to near certainty.

In contrast, X's characteristics correspond to an odds multiplier of 900, so the odds that X is a terrorist go from 1 to 10,000, to 900 to 10,000. The probability that X is a terrorist thus goes from 0.01% to about 9%: from "highly unlikely" to "rather unlikely." Our initial intuition that X is more likely to be a terrorist than Z might be incorrect. The crucial difference is that Z is drawn from a group with a higher base rate than X, so even a modest odds multiplier greatly increases the likelihood that Z is a terrorist; in contrast, X's group has such a low base rate that even a sizable odds multiplier will not substantially change the probabilities.

The Promise and Peril

It follows that if our profile is to be effective, we should first identify a group with a high base rate, then apply our profile to members of that group. But how can we identify a group with a high base rate? The obvious strategy is to use the odds multipliers themselves.

The fundamental assumption of data mining is that dozens, if not hundreds, of characteristics can be analyzed. As long as we take into account both matches and non-matches, we could simply apply all elements of our profile and produce a list of persons with a very high odds multiplier. It's not entirely clear where such a search would stand constitutionally. We might ask whether it constitutes a reasonable search. One of the key factors that determines whether a search is reasonable involves how much it violates an expectation of privacy. In particular, the greater the expectation of privacy, or the more invasive and inconveniencing a search, the more likely a warrantless search will be deemed unreasonable.

Data mining is not at all inconveniencing: indeed, the subjects might never be aware that it occurred. As for being invasive, that is less clear. One of the more pragmatic jurists and legal theorists, Judge Richard Posner of the Seventh Circuit Court of Appeals, argues that data mining per se is not an invasion of privacy, since the initial sift of records is done by computer, and an invasion of privacy can only occur when and if a human being examines the records flagged for further investigation.

This is a point of view worth considering: at what point does the invasion of privacy occur? For example, consider a person's medical records, for which there is some expectation of privacy. Clearly the records exist somewhere, and the existence of the record is *not* an invasion of privacy. If there is no external mark indicating the person's name, is there a privacy violation when their file is moved to the desk of a police officer on vacation? Or does the invasion of privacy not occur until the officer returns and opens up the file?

Illinois v. Gates (1983) provides a potentially useful precedent. In this case, Bloomingdale police received an anonymous letter identifying Sue and Lance Gates as drug dealers, accusing them of buying drugs in Florida to be driven to Chicago. Independent investigation corroborated certain aspects of the letter, and a search warrant was obtained for the Gates's home, where a large stash of marijuana was discovered. The U.S. Supreme Court ultimately upheld the search, holding that while the anonymous letter by itself would not provide probable cause for a warrant, the subsequent investigation did.

By comparison, a data mining operation might be deemed equivalent to an anonymous letter pointing to a person in whom the authorities might be interested. Following *Illinois v. Gates*, this could be used to start an investigation based on information that could be obtained without a warrant. If corroborating evidence appeared, a warrant could be obtained and further action could be taken; otherwise, the investigation should be abandoned (and from a pragmatic point of view, the resources could be reallocated to investigations more likely to be productive).

There is a risk (borne out by recent allegations) that datasets obtained this way can be misused. But since misuse is not sanctioned by legal warrants, perpetrators are open to civil and criminal charges, and the fact that the datasets *could* be misused is only minimally relevant to the question of whether their *existence* is unconstitutional. By analogy, the fact that a federal agent *could* use a firearm to break the law should not cause us to deny them access to firearms.

The Cold Equations

This purely quantitative approach may be the best possible guardian of our civil liberties. As long as we rely on human beings to evaluate matches and non-matches, we must rely on their judgment regarding the significance of a match and non-match. This judgment can be impaired by a number of factors: an appealing narrative, a desire to catch a suspect, or less savory aspects of human nature. Moreover, what appears to one person as a flagrant violation of civil liberties will invariably be viewed by another as a reasonable compromise between constitutional rights and security.

In contrast, the cold equations are completely objective. Matches and non-matches are treated as they should be, and not weighted inappropriately based on personal experience, allowing the pool of persons of interest to be effectively reduced. From a civil liberties point of view, this limits the number of innocent people caught up in an investigation. From a security point of view, this allows limited law enforcement resources to be spent in the most efficient way possible. From both points of view, further increasing the efficiency of the search algorithm is desirable. The issue is not *whether* data mining and similar strategies should be used, but rather increasing and improving the efficiency of these strategies so they involve a lesser risk to civil liberties.

"The Man of Statistics"

No person shall be held to answer for a capital, or otherwise infamous crime, unless on a presentment or indictment of a Grand Jury . . . nor shall any person be subject for the same offense to be twice put in jeopardy of life or limb.   —U.S. CONSTITUTION, AMENDMENT V

Clarence Norris was one of nine black teenagers accused of raping Victoria Price and Ruby Bates on a train on March 25, 1931. The U.S. Supreme Court began hearing the case of *Norris v. Alabama* on February 18, 1935. Norris had been indicted by a grand jury, found guilty by a petit (trial) jury, and sentenced to death. But there was one disquieting factor that cast doubt on the validity of the process: Norris was black, while Price, Bates, and all members of both the grand and petit jury were white.

Through a bizarre concatenation of events, the American Communist Party (ACP) picked up the cause of the Scottsboro boys, and recruited New York lawyer Samuel Leibowitz to represent the defendants, beginning with Haywood Patterson. With more competent counsel, the defects of the case against the Scottsboro boys became evident. The physical evidence suggested consensual sex days earlier, and not rape on March 25, 1931. The only eyewitness claimed to have seen the women in dresses, though in fact they were in overalls. Ruby Bates recanted her accusation. The dubious nature of the evidence, combined with the fact that a black youth accused of raping a white woman was represented by a Jewish lawyer from New York hired by a branch of the American Communist Party, caused Judge James Horton to remind the jurors that neither race, nor lawyers, nor state lines should play a role in their decision. But after just one day of deliberation, the jury found Patterson guilty. Leibowitz filed motion for a new trial; Judge Horton stunned everyone on June 22, 1933 when he granted it. This courageous decision would cost Horton re-election.

The prosecution changed the trial venue to Decatur, in Morgan County, with Judge William Callahan presiding. Once again, Patterson was found guilty, as was Clarence Norris; once again both were sentenced to death. However, anticipating a lengthy appeals process, Callahan postponed the trials of the remaining seven (who would remain incarcerated until trial, even though they had not yet been convicted of any crime).

The appeal centered around a crucial fact. According to the 1930 census, Jackson County (where the indictments were handed down) had a population of 36,881, of whom 2,688 were black; this is roughly 7% of the population. Yet a number of local citizens, as well as two court clerks, two jury commissioners, and a court reporter who had not missed a case in 24 years, testified that they had never seen a black man called for jury service; one witness went so far as to aver that no black had ever served on a grand or petit jury in the history of the county. The situation was similar in Morgan County, where the trials were held: the 1930 census gave Morgan County a population of 46,176, of whom 8,311 (about 18%) were black, yet in living memory, not one had ever served on a jury.

The jury commissioner for Morgan County gave a simple explanation for the absence of blacks from the grand juries:

> I do not know of any negro in Morgan County over twenty-one and under sixty-five who is generally reputed to be honest and intelligent and who is esteemed in the community for his integrity, good character and sound judgment, who is not an habitual drunkard, who isn't afflicted with a permanent disease or physical weakness which would render him unfit to discharge the duties of a juror, and who can read English, and who has never been convicted of a crime involving moral turpitude. (294 U.S. 587, pp. 598–599)

To counter this astounding claim that an entire population group was unfit for jury duty, Leibowitz produced a list of African Americans who were teachers or members of school boards; college graduates; business owners; and others of similar stature. He also pointed out that several African Americans had been called to serve as jurors in *federal* courts. The evidence persuaded the Supreme Court and, on April 1, 1935, it reversed the Alabama State Supreme Court's decision to uphold the conviction.

Unfortunately, the state was free to retry the case. A single African American, Creed Conyer, sat on a new grand jury that heard the case. However, since the grand jury only needed a two-thirds vote to indict, it was able to hand down new indictments on November 13, 1935. Once again, after a short trial presided over by Judge Callahan, Norris was found guilty and sentenced to death, though the governor (Bibb Graves) reduced the death sentence to life in prison; Norris was paroled in 1946. In 1976, Alabama governor George Wallace issued an official pardon, bringing closure to one of the more shameful events in American history.

The Court of Numbers

The argument made by the jury commissioner in *Norris* had been the standard tactic to defuse charges of jury discrimination for 50 years: no blacks appeared on juries because no blacks were suitable for jury duty. Ironically, this defense worked because race-based selection of jurors was (and remains) a violation of federal law. If we claim that jury commissioners used race to determine whether to include a person on a jury list, we place them in the role of being the defendant in a criminal trial. However, defendants in a criminal trial have a presumption of innocence: we must assume they are innocent, until and unless proven guilty. Indeed, in 1986 the Supreme Court observed that it could find no evidence that anyone had ever been prosecuted for a violation of this law.

The problem is this. Suppose the state selected jurors on the basis of race. This would make it easy to produce an all-white jury. But if the state completely ignored race in the selection of jurors, it could not *prevent* the formation of an all-white jury. This means we need a way to distinguish between an all-white jury formed by deliberate exclusion of blacks, and an all-white jury formed by completely ignoring race in the selection of jurors. Oliver Wendell Holmes, Jr. alluded to a solution this problem in 1897, five years before his appointment to the Supreme Court. Holmes opined: "For the rational study of the law the black letter man may be the man of the present, but the man of the future is the man of statistics and the master of economics."

The statistical approach to deciding which of two explanations of an observation we should accept is known as *statistical decision theory* (SDT), and bears many similarities to how our legal system works; in fact, the most common way to introduce it in a mathematics class is in the context of a criminal trial.

It is convenient to think of a criminal trial as starting with a default assumption, namely that the defendant is innocent. The state must then provide sufficient evidence to cause the jury to reject this assumption and accept the alternate conclusion of guilt. Note that under this system, a defendant is *never* proven innocent; he or she is merely not proven guilty.[1]

In statistics, the default assumption is (for reasons discussed later) known as the null hypothesis. If we have enough evidence, we can reject the null hypothesis and accept the alternate hypothesis. Otherwise, we fail to reject the null hypothesis. There is one key difference between statistics and the law. Since we never accept the null hypothesis, we could in principle collect more evidence and reject the null hypothesis at a later date. In contrast, the 5th Amendment prohibits trying a person twice for the same crime. Thus the law makes failure to reject the null hypothesis equivalent to accepting the null hypothesis.[2]

While we hope that our trial system makes the correct decision in every case, we must accept that it will occasionally make mistakes. There are two types of errors. First, we might convict someone who is innocent: a wrongful conviction. In statistical terms, we have incorrectly rejected the null hypothesis. Statisticians call this a *Type I error*, but we will use a more evocative term drawn from signal theory: a *false alarm*. The other possibility is that the jury will acquit someone who is guilty: a wrongful acquittal. Statisticians call this a *Type II error*; again, we will use the term from signal theory: a *missed alarm*.

The existence of the two types of errors is important, because while our goal is to minimize the likelihood of making *any* mistake, the universe rarely gives us something for nothing. In this case, decreasing the frequency of one type of error increases the frequency of the other type.

For example, consider a criminal justice system where being in the vicinity of a crime is sufficient for a conviction. If we arrested and charged everyone in the vicinity of the crime, we would have a very low wrongful acquittal rate: the perpetrator will almost certainly be convicted. At the same time, many of those convicted would be innocent, so the wrongful conviction rate will be very high. This system would have a very low miss rate, but a high false alarm rate.

---

1. The Scottish legal system recognizes this distinction: the jury may convict or acquit a defendant, but it has the option of returning a verdict of "not proven."

2. The Scottish "not proven" verdict is also treated as an acquittal; defendants so judged cannot be retried.

On the other hand, consider the opposite extreme, where the only way to obtain a conviction is to have incontrovertible evidence, say in the form of a video of the crime that clearly shows the perpetrator and victim, with the perpetrator self-identifying. Under such a system, very few innocent persons would be convicted. On the other hand, there would be many crimes for which no such video existed; thus there would be a great many wrongful acquittals. In practice, such a system would mean that most cases never went to trial. These would be wrongful acquittals by the *system*, if not necessarily by the *jury*. This system would give us a very low false alarm rate, but a very high miss rate.

Clearly we want to balance too high a false alarm rate and too high a miss rate. The law operates by setting a standard, then letting future generations decide whether the false alarm and miss rates are palatable. But with mathematics, we need not wait for the judgment of history: we can determine the error rate directly.

To see how this works, let's consider a simpler problem: deciding whether a coin is fair or not. To do this, we must first define what we mean by a "fair coin." Most people would say that a fair coin lands heads or tails with equal frequency. However, there is a hidden assumption in this notion of fairness, which can be illustrated as follows: Suppose you saw a person flipping a coin, and observed that the coin consistently alternated between heads and tails. By the preceding definition, the coin is fair—but would you wager $100 on the result of the next flip?

This suggests that a better definition of a fair coin is one that lands heads or tails with equal frequency in the long run. Using this sense of what constitutes a fair coin, most people would try to determine whether a coin is fair or unfair by flipping it repeatedly and seeing how large a disparity accumulates in the short run. Since we defined a fair coin as one that landed heads and tails with equal frequency in the long run, this means that if we flipped this coin many times, we should see it land heads 50% of the time and tails the remaining 50% of the time. Thus we say that the probability the coin lands heads is 50%; likewise, the probability the coin lands tails is also 50%.

Now consider our problem. Suppose we flip a coin ten times and observe it land heads eight times. Let us rule that "chance and accident" could not have caused this, and conclude that the coin is unfair. The law is based on precedent, so making a decision in this particular case creates a standard that is usable for all future cases. We might word our decision as follows:

Landing heads eight times out of ten flips is prima facie evidence that the coin is not fair. Legal philosophy mandates that no single piece of evidence is conclusive; rather, it is the totality of all evidence that persuades or fails to persuade. We will not say that the coin *is* unfair, but rather that we have prima facie evidence to conclude that the coin is unfair. The distinction will become important later. For now, we'll treat prima facie evidence as conclusive.

But precedent is a guide, not a straightjacket, and too narrow an interpretation of precedent makes it useless. In this case, we should ask why eight heads in ten flips suggest an unfair coin. Most people would justify such a decision on the basis that it was unlikely for a fair coin to produce so great a disparity between the expected and observed numbers of heads. To determine how often this occurs, mathematicians use what is known as a *binomial probability*: in this case, the probability that a fair coin lands heads eight times in ten flips is about 0.04395.

Now consider the implication of the decision "Landing heads eight out of ten times is too improbable to occur by chance." This means that we should consider any event with probability 0.04395 or less too improbable to occur by chance. Consequently, if we observe any event with a probability of 0.04395 or less, we can use our observation as prima facie evidence that the coin is unfair.

Let's calculate the probability that a fair coin lands heads 0, 1, 2, 3 . . . 8, 9, 10 times in 10 flips. If we compute these probabilities, we see that the probability that a coin lands heads 9 times in 10 flips is 0.00977, so if we observed this occurrence, we should also conclude it too improbable to occur by chance. Again, the probability that a coin lands heads 10 times in 10 flips is 0.00098, so observing this event would cause us to doubt the fairness of the coin. By similar computations, we'd find that observing 0, 1, or 2 heads in 10 flips would cause us to doubt the fairness of the coin.

This means that if we decide that observing the coin land heads eight times in ten flips provides sufficient evidence for deciding the coin is unfair, then we have established the rule that landing heads eight or more times in ten flips, or two or fewer times in ten flips, is prima facie evidence that the coin is unfair.

In law, we'd set a standard like the preceding, then use it for some time and allow the court of public opinion to decide whether or not the standard produced an acceptable balance between the false alarm and miss rates. But if our standard is flawed, it follows that we will be using a flawed standard for some time, which harms all those who are judged by it. Instead, we can use mathematics and the court of numbers to predict how well the standard will perform before we use it even once.

To analyze this standard mathematically, we begin by trying to determine the false alarm rate: the probability of incorrectly rejecting the null hypothesis. Since there are two hypotheses (that the coin is fair, or that the coin is unfair), and coins do not have a presumption of fairness, it might not be clear which one we should take as the null hypothesis. However, notice that the first hypothesis assumes that the coin lands heads with probability 50%. Because it gives us a usable probability, we will make it our null hypothesis, "null" because it assumes that nothing but known random factors affect the outcome of the coin toss.

Our rule sounds a false alarm whenever a *fair* coin lands heads eight or more times, or two or fewer times. The probability works out to be about 11%. This value requires careful interpretation. It does *not* mean that this method of deciding whether a coin is fair will be incorrect 11% of the time. Rather, it means that if you have (unknown to you) a *fair* coin, this standard will cause you to make the incorrect decision 11% of the time.

To understand the distinction, suppose we had a large group of people, and gave each member the task of deciding whether the coin in front of them is fair or not, using the above rule. If we placed a fair coin before each of them, about 11% of them would report the coin unfair, and every one of these declarations is a false alarm. Since we can't have more than 100% of the coins being fair, it follows that the maximum false alarm rate will be 11%. Note that this does *not* mean that less than 11% of the declarations that the coin is unfair will be false alarms. In the case where everyone has a fair coin, 100% of the declarations that the coin is unfair will be false alarms.

What about the miss rate? We have a miss whenever we declare an unfair coin to be fair; this would occur if an unfair coin landed heads between three and seven times. Unfortunately, the alternate hypothesis gives us no usable probability, so we cannot determine the miss rate. This is true in general: unless both hypotheses give us usable probabilities, the miss rate cannot be determined. All we can do is remember that the higher the false alarm rate, the lower the miss rate, and vice versa.

The use of mathematical probability allows us to extend our legal standard as follows. Again, the law is based on precedent, so when we agreed that eight heads in ten flips constituted prima facie evidence of unfairness, we also agreed that a false alarm rate of less than 11% was acceptable.

The introduction of the acceptable false alarm rate is critical, for it allows us to consider a broader range of possibilities. For example, suppose we are confronted with some evidence, say that a coin landed heads 187 times out of 336 tosses. We do not have a situation where we have flipped the coin ten times. Nor can we simply look at ten of the results, since it is unreasonable to discard 97% of our observations without some rationale for which observations are kept and which are discarded. Thus we cannot apply our original rule, since it assumes we have flipped the coin exactly ten times.

It's tempting to use our original decision that 8 heads in 10 flips is too improbable to occur by chance. In this case, the probability that a fair coin lands heads 187 times in 336 flips is 0.00508, which is less than the probability that a fair coin lands heads 8 times in 10 flips. Should we then conclude, on the basis of this probability, that the coin is unfair?

The problem is that as the number of flips increases, the probability of observing any specific number of heads falls. Indeed, suppose we observed exactly 168 heads in 336 flips—in other words, the coin landed heads exactly half the time. No sensible person would conclude that the coin is unfair—yet the probability that this event occurs is only 0.04350. Thus if we base our decision on the improbability of the specific event we observe, we would have to conclude that such a coin is unfair!

To avoid this counterintuitive situation, we can compare the false alarm rates. We accepted a rule that gave us a false alarm rate of 11% when flipping a coin ten times; thus we should accept a rule that gives us a false alarm rate of 11% when flipping a coin 336 times. With some effort, we find that we obtain this false alarm rate with the following rule: conclude the coin is unfair if you see 183 or more, or 153 or fewer, heads in 336 flips.

The computations needed to translate a false alarm rate into a rule for a particular case are not difficult, though they are tedious and time-consuming. One way to speed this process emerges as follows. Suppose we gave a large number of people a fair coin, and directed them to flip it 336 times and record the number of heads they observed. Unsurprisingly, these numbers are randomly distributed. For example, if we gave fair coins to ten persons, they might report the number of heads observed as:

187, 170, 163, 181, 184, 179, 165, 177, 159, 161

In other words, the first experimenter observed 187 heads out of 336 flips; the second observed 170 heads in 336 flips; the third 163 heads in 336 flips; and so on.

The exact probability of any of these outcomes can be found using the binomial probability distribution, but when the numbers are even modestly large, as in this case, we can use the *normal approximation to the binomial distribution* to compute them. For example, the probability that a fair coin lands heads 187 times in 336 flips is about 0.005082, while the normal approximation to the binomial gives us 0.005085.

We can think of this loss in accuracy as the price of using the normal approximation to the binomial distribution. What we buy is a very easy way to compute the probability of a range of outcomes. For example, if we wanted to find the probability that a fair coin landed heads 183 or more times in 336 flips, using the binomial distribution would require us to compute the probability that the coin landed heads 183 times; 184 times; 185 times; and so on, up to 336 times; then add these probabilities together. In contrast, the normal distribution allows us to use the tools of calculus to find the probability much more quickly.

The normal distribution requires two key parameters: the mean and the standard deviation. There is a simple formula for computing the mean and standard deviation when using the normal distribution to approximate the binomial: in this case, our mean would be 168 heads and our standard deviation would be about 9.1652.

We can think of the normal distribution as a machine that produces a value we might have obtained had we run the corresponding experiment. Rather than give fair coins to ten different experimenters and have them flip the coin 336 times, we could instead run the "normal distribution machine" repeatedly. Every time we ran the machine, it would output a number between 0 and 336, which we could interpret as the number of heads an experimenter would observe if he or she flipped a fair coin 336 times.

As with any individual coin toss, we cannot predict what the normal distribution machine will produce on any given run. However, if we run it a great many times, we can confidently say:

1. About 68% of the time, we will obtain a value within 1 standard deviation of the mean.
2. About 95% of the time, we will obtain a value within 2 standard deviations of the mean.
3. About 99.7% of the time, we will obtain a value within 3 standard deviations of the mean.

This is sometimes known as the 1-2-3 rule. If this looks familiar—it is. Confidence intervals are based on the normal distribution.

For example, suppose we set our normal distribution machine to have a mean of 168 and a standard deviation of 9.1652. This corresponds to the experiment of flipping a fair coin 336 times and recording the number of heads. If we run the normal distribution machine repeatedly, we would find (in accordance with the 1-2-3 rule) that about 95% of the numbers produced are within two standard deviations of the mean: in other words, between $168 - 2 \times 9.1652 \approx 149.67$ and $168 + 2 \times 9.1652 \approx 186.33$. Since our normal distribution machine simulates what happens when we run the actual experiment, we can translate this result into another: if we asked a large number of people to flip a fair coin 336 times, about 95% of them would see between 150 and 186 heads. Moreover, about 5% of them would see more than 186, or fewer than 150, heads on 336 flips.

Now consider the rule: if you see more than 186, or fewer than 150, heads on 336 flips, conclude that the coin is unfair. By the preceding, a fair coin would be judged unfair about 5% of the time; thus our false alarm rate for this rule is 5%. The 5% false alarm rate is known as the *level of significance*. While we could in theory use any level of significance we want, commonly used rates include 5%, 1%, and 0.1%.

Note that our threshold numbers (186 and 150) were respectively two standard deviations above the mean, and two standard deviations below the mean. Thus we could rephrase our rule as follows: if the number of heads is more than two standard deviations from the mean, conclude that the coin is unfair. The advantage is that this statement of the rule gives us the same false alarm rate—5%—regardless of the actual mean or standard deviation. Moreover, since we can calculate the mean and standard deviation for any number of flips, we can judge the facts before us, without having to obtain additional evidence or decide which evidence to exclude.

For example, suppose we flip the coin 36 times and observe 10 heads and 26 tails. This situation can be modeled using a normal distribution with mean 18 and standard deviation 3. The rule "If the number of heads is more than two standard deviations from the mean, conclude that the coin is unfair" becomes "If the number of heads is more than $18 + 2 \times 3 = 24$ or less than $18 - 2 \times 3 = 12$, then conclude the coin is unfair." In the absence of mathematics, we might wonder whether we can apply a rule (based on 336 coin tosses) to the situation when we only toss the coin 36 times; we might consider the evidence insufficient, and run more experiments to accumulate more data. With mathematics, we can translate the rule for 336 coin tosses into an equivalent one for 36 coin tosses using the data on hand.

"Suspect to a Social Scientist"

In March 1972, Rodrigo Partida had been convicted on a charge of burglary with intent to rape. But even though Mexican Americans made up 79.1% of the population of Hidalgo County, Texas, they made up only about 39% of those summoned for grand jury service over an 11-year period. Partida appealed, charging discriminatory selection of grand jurors.

Not everyone is eligible for jury duty. However, the evidence suggested that, while some Mexican Americans would be ineligible for jury service, most would be. Citing the disparity between the percentage of Mexican Americans who served on juries (about 39%) and the percentage of eligible jurors who were Mexican American (unknown, but not much less than 79%), the U.S. Supreme Court ruled in *Castaneda v. Partida* (1977) that jury discrimination had occurred. Consequently, Partida's conviction was overturned.

Ostensibly the decision by the Supreme Court was based on the substantial difference between the expected and observed percentage of Mexican American jurors. But Justice Harry Blackmun, writing the majority opinion, made an oblique reference to the 1-2-3 rule and statistical decision theory: "As a general rule for such large samples, if the difference between the expected value and the observed number is greater than two or three standard deviations, then the hypothesis that the jury drawing was random would be suspect to a social scientist" (430 U.S. 482, p. 496).

The null hypothesis is that the summoning of grand jurors was being done without regard to race or ethnicity, which would give 79.1% as the probability that a person summoned for jury duty was Mexican American. Over the 11-year period, 870 persons were summoned for grand jury service, so the number of Mexican Americans we would expect to see would be 79.1% of 870, or about 688, which we take as our mean; our standard deviation works out to be about 12.

Suppose we drew up several 870-person lists of persons to serve as grand jurors. By the 1-2-3 rule, about 99.7% of our lists would include between 652 and 724 Mexican Americans. It would be very unlikely for a list to contain fewer than 652 or more than 724 Mexican Americans. But in fact, Hidalgo County, Texas only summoned 339 Mexican Americans during that time period. The unlikelihood of this observation occurring by "chance or accident alone" suggests that some other factor was in operation: we reject the null hypothesis (that the jurors were being selected without regard to race or ethnicity) and hold that there is prima facie evidence that jury discrimination has occurred.

Two or Three?

One problem with the *Castaneda* decision is that the Court was deliberately vague: a disparity of "two or three" standard deviations would cause a social scientist to find the claim of randomness to be suspect. This raises the question of how much of a disparity is necessary to establish a prima facie case of race-based discrimination.

To understand the significance of this question, let us consider two statisticians trying to determine whether a coin is fair or not. Both statisticians use the observation that the coin has landed heads 63 times in 100 flips, but one of them insists on a disparity of three standard deviations, while the other would accept a disparity of only two.

Both statisticians would analyze the problem using the normal distribution with mean 50 and standard deviation 5; both would identify the same disparity (13 more heads than expected). They differ in their interpretation of the observation. The first statistician, observing that the disparity is less than three standard deviations, would rule that the evidence is *not statistically significant*. The second statistician, observing that the disparity is more than two standard deviations, would disagree and say that the evidence is *statistically significant*.

Conflicting conclusions like these seem to support the idea that statistics can be made to say anything you want them to. This is true—provided you hide part of the statistics. In this case, what is not indicated is the level of significance. The first statistician, by insisting on a three standard deviation disparity, is also insisting on a false alarm rate of no more than 0.3%. Thus their conclusion, when not reduced to a bumper sticker or a sound bite, is that the evidence is not statistically significant *at the 0.3% level of significance*. On the other hand, the second statistician, who accepted a two standard deviation disparity as sufficient, also accepted a miss rate of as much as 5%. Their complete conclusion is that the evidence is statistically significant *at the 5% level of significance*.

Both statisticians would agree that the probability that a fair coin landed heads 63 times out of 100 is less than 5%. The difference is that one statistician is willing to accept a 5% false alarm rate, while the other is not. It is tempting to side with the first statistician and insist on a low false alarm rate. But keep in mind that the lower the false alarm rate, the higher the miss rate. We must decide the proper balance between the false alarm and miss rates. To that end, we must consider the consequences of an erroneous decision.

In the case of jury discrimination, the null hypothesis is that no discrimination exists. A false alarm would result in a declaration that jury discrimination existed when it actually did not. This could require the retrial of every case heard by the jurisdiction over the period in question, at a considerable cost in time and money; it is even possible that some persons, previously convicted, would be acquitted. On the other hand, a miss would cause us to conclude that no jury discrimination existed when in fact it does. As a result, a discriminatory juror selection system would remain in place. This risks placing defendants before biased juries, which strikes at the heart of our notion of justice.

If we feel the consequences of a false alarm (retrying cases unnecessarily) are too great, then we should require a disparity of three standard deviations. But if we feel the consequences of a miss (allowing a discriminatory system of selecting jurors to continue) are too great, then a disparity of two standard deviations should suffice for statistical significance.

The decision of how much disparity sufficed would be made by the Supreme Court in *Vasquez v. Hillery* (1986). In 1962, Booker T. Hillery, an African American ranch hand, was indicted for the murder of 15-year-old Marlene Miller of Kings County, California. Hillery was subsequently convicted and sentenced to death. Of the 210 grand jurors selected (by the same judge) between 1956 and 1962, none were African American, though about 4.6% of the population was, suggesting jury discrimination; Hillery appealed.

In *Hillery v. Pulley* (1983), heard by the District Court, Hillery introduced a standard deviation analysis. The difference between the expected number (4.6% of 210, or about ten) and the observed number (zero) was just over three standard deviations. This exceeded Castaneda's "two or three" standard deviations, and Judge Karlton of the District Court ruled in favor of Hillery: "I conclude that statistical evidence can, and in this case does, provide a basis for an inference of discriminatory intent" (563 F. Supp. 1228, p. 1246). The U.S. Court of Appeals for the Ninth Circuit agreed, and in *Vasquez v. Hillery* (1986), the Supreme Court upheld the District Court's ruling and established that disparities of greater than three standard deviations could be taken as prima facie evidence of jury discrimination.

Unfortunately, while the Supreme Court has established clear guidelines for what constitutes prima facie evidence of unconstitutional grand juror selection, it has consistently failed to do so for petit juror selection; moreover, lower courts have consistently denied the acceptability of the statistical methods used in *Castaneda* and *Vasquez*. Indeed, as we shall see in the next section, the only guidance given by the Supreme Court is a horrifically flawed standard that almost seems designed to perpetuate, not prevent, discrimination against minorities.

Despair over Disparity

> In all criminal prosecutions, the accused shall enjoy the right to a
> speedy and public trial, by an impartial jury of the State and district
> wherein the crime shall have been committed.
>
> —U.S. CONSTITUTION, AMENDMENT VI

On November 7, 1991, Diapolis Smith shot Christopher Rumbly in the chest
during a bar fight in Grand Rapids, Michigan. Smith was convicted of second-
degree murder and sentenced to life imprisonment. His case was decided
by an all-white jury drawn from an all-white venire, so he appealed on the
grounds of jury discrimination.

The Michigan Court of Appeals agreed that race-based discrimination
had occurred in the jury selection process, and granted Smith a new trial.
But the state appealed, and the Michigan Supreme Court held that insufficient evidence had been presented to support Smith's contention. The Sixth
Circuit Court of Appeals disagreed, and again concluded that race-based
discrimination had occurred. The state appealed again, and on January 20,
2010, the U.S. Supreme Court heard *Berghuis v. Smith* and confronted more
than a century's worth of decisions regarding the composition of a trial
("petit") jury.

An Impartial Jury

While state and federal laws set requirements for jury service, not everyone
who qualifies is called. In most jurisdictions, a group of potential jurors is summoned; those who answer are known as the *venire* (from the Latin word meaning "to come"). Generally, the venire consists of far more people than necessary, so jurors go through an interview process known as *voir dire* (from an

Anglo-Norman phrase meaning "to speak the truth").[1] This process eliminates jurors who might be biased, and gradually reduces the venire to the jurors, plus some alternates, who actually sit on the *jury panel* and decide the case.

Popular culture equates the constitutionally mandated impartial jury with a "jury of one's peers." This was one of the rights granted to the nobility of England in the *Magna Carta* (1215), and it appears in Blackstone's *Commentaries on the Laws of England* (1765). The phrase does not appear in the Constitution, and has no legal status in American law.

Moreover, the "jury of one's peers" gives rise to more problems than it solves. For example, consider a case with a black defendant. It might seem reasonable to try black defendants using an all-black jury. While this might be appealing, there are several problems with aligning the race of the jurors with the race of the defendant. First, it is a variation of the "separate but equal" doctrine of *Plessy v. Ferguson* (1896) that perpetuated segregation in the United States. Next, what if the race of the defendant and victim are different? If we hold that a black defendant can be tried by an all-black jury, even if the defendant is white, the equal protection clause of the 14th Amendment would require that we grant a *white* defendant a similar right to an all-white jury—even if the victim was black.

Thus, beginning with *Strauder v. West Virginia* (1879), the Supreme Court has consistently denied the right of a defendant to a jury of any particular composition. It was not until *Glasser v. United States* (1942) that the Supreme Court addressed the issue of petit jury composition.

Daniel D. Glasser and codefendant Norton I. Kretske were assistant U.S. attorneys for the Northern District of Illinois. They were responsible for prosecuting liquor-related tax evasion, but had been accused and convicted of taking bribes. They were tried by a jury consisting of six men and six women. However, all the women on the venire were members of the Illinois League of Women Voters. Justice Frank Murphy, writing the opinion of the majority, established what has become known as the *fair cross-section requirement*: a jury should be a "body truly representative of the community." Unfortunately, the Supreme Court went on to echo the viewpoint that

---

1. In modern French, "voir" is "to see," which leads to the erroneous etymology of *voir dire* as "to see, to speak." However, the term originated from medieval French, where "voir" means "the truth."

perpetuated race-based discrimination for a half century: the total exclusion of a group is insufficient to prove deliberate discrimination against the group. Thus the Supreme Court upheld Kretske's conviction (though overturned Glasser's on other grounds). Fortunately, it only took six years before the Supreme Court changed its mind on how to evaluate petit jury composition. The decision would put it on a collision course with the states.

Eddie Patton was a black man charged with murdering Jim Meadows, a white man. An all-white grand jury indicted Patton; an all-white petit jury convicted him; and he was sentenced to death by electrocution. He appealed, charging discriminatory selection of the jury venire: blacks made up about 36% of the adult population of Lauderdale County, Mississippi, where the case was tried, so he argued that the complete absence of blacks from the grand and petit jury venires in his case (and for all cases stretching back 30 years) had to be caused by race-based selection of jurors. The Mississippi State Supreme Court denied the appeal, but future Supreme Court Justice Thurgood Marshall took up Patton's case, arguing it before the Supreme Court in *Patton v. Mississippi* (1947) on November 21, 1947.

Marshall argued that the exclusion of blacks from the grand and petit juries violated the equal protection clause of the 14th Amendment. The state of Mississippi countered by pointing out that very few blacks met the eligibility requirements, which included being a registered voter. As a result, there were only about 25 blacks eligible for jury service in the county, in contrast to about 6,000 eligible whites. Marshall countered with evidence that blacks had, by and large, been prevented from registering to vote.

Voter discrimination would become a critical issue during the Civil Rights era, and the Supreme Court had the opportunity to raise the question in *Patton*. But it had to move carefully. Just one year before, in *Morgan v. Commonwealth of Virginia* (1946), it overturned a Virginia law mandating the separation of black and white passengers on buses.[2] This set the stage for an

---

2. On July 14, 1944, Irene Morgan had been arrested for refusing to give up her seat to a white passenger. Marshall, who also argued this case before the Supreme Court, won through the use of an unexpected argument: the state law placed an undue burden on interstate commerce. Marshall won his case, but the reliance on the interstate commerce argument meant that segregation on *city* buses was still permissible, leading to Rosa Parks's more famous ride, 11 years later.

epic conflict with state-sponsored discrimination, and moving too quickly too early could have catastrophic effects.

Thus, the Supreme Court declined to consider *why* blacks were underrepresented on jury venires. Rather, it ruled that since the *system* produced a "long and continued exclusion" of blacks from juries, the indictments and convictions produced could not stand. The Supreme Court overturned Patton's conviction.

Mississippi's response was to retry, reconvict, and resentence Patton, this time by a petit jury that included one black. On appeal, again on the basis of discriminatory selection of jurors, the Supreme Court of Mississippi attributed the small number of blacks on the venire to "lethargy which had theretofore brought the comfort of an absolution from civil duties" (563 F. Supp. 1228, p. 1246). Concluding that the jurors had been selected in compliance with the Supreme Court's directives, the conviction was upheld, and Patton was executed on January 12, 1950.

White and Yellow

Mississippi ended the "complete and long-continued exclusion" of blacks from juries by including one black on the jury that tried and convicted Patton. That it obeyed the letter of the law and not the spirit is beside the point: the Supreme Court bears the responsibility for issuing decisions where the letter and the spirit coincide. Mathematics offers a pathway—if the courts are willing to take it.

The first steps in this direction were taken in *Avery v. Georgia* (1953). James Avery, a black man, was convicted of raping a white woman. An all-white jury drawn from an all-white venire convicted him and he was sentenced to death.

The case was tried in Fulton County, Georgia, which used a seemingly neutral system of producing a venire: the names of eligible jurors were put on slips, which were put into a box. A judge then drew the veniremen required for a week's worth of trials; the slips would be allocated among the criminal and civil trials on the docket. Such a system would seem to preclude any possibility of discrimination on the basis of race. However, there was a peculiar feature: white jurors had their names on white slips, while black jurors had their names on yellow slips.

For Avery's trial, 60 veniremen were drawn. Since 5% of the eligible jurors were black, one would expect such a drawing to produce a venire with about three blacks. Yet the venire drawn was all-white. Justice Felix Frankfurter opined that "The mind of justice, not merely its eyes, would have to be blind to attribute such an occurrence to mere fortuity" (345 U.S. 559, p. 564). On May 25, 1953, the Supreme Court reversed the decision by the Supreme Court of Georgia and overturned Avery's conviction.

As in *Patton*, the state was free to re-prosecute. The second time around, the issue of jury discrimination was avoided by eliminating the jury: on May 1, 1954, Avery accepted a plea-bargain sentence of 20 years in prison, rather than face a trial and potential death sentence.

The *Avery* case was decided before the Supreme Court began accepting statistical arguments. Ironically, a statistical analysis might have produced a different result. If 60 veniremen were drawn from a population that was 5% black, one would expect to see three black veniremen; we actually observed zero. Rejecting the null hypothesis that jurors were selected without regard to race translates into a false alarm rate of about one in eight—rather higher than the highest commonly accepted false alarm rate, and casting doubt on Frankfurter's conclusion that "mere fortuity" could not produce the observed results.

The problem the Supreme Court faced might be phrased as follows. *Glasser, Patton,* and *Avery* established that the petit jury venire should reflect a fair cross-section of the community. But we would not expect an exact match. How large a disparity must accumulate before we have a prima facie case of discrimination? The Supreme Court addressed this issue in *Swain v. Alabama* (1965).

Robert Swain, an African American, was accused of raping a white woman; a jury in Talladega County, Alabama convicted him and sentenced him to death. At the time, jurors in Alabama had to be male and over the age of twenty-one. About 26% of the eligible jurors in Talladega County were black, but between 1953 and 1963, only about 10–15% of the grand and petit jury venires were black; and while blacks had served on grand juries, none had served on a petit jury since before 1950.

Chief Justice Warren, as well as Justices Arthur Goldberg and William Douglas, agreed that the disparity between the percentage of eligible blacks and the percentage of blacks on the venire constituted prima facie evidence of illegal jury discrimination. Unfortunately, the other six justices did not. Thus on March 8, 1965, the Court upheld Swain's conviction. Swain's death sentence was eventually commuted to a lengthy prison sentence, and he would be released in 1996.

The *Swain* ruling established what is now known as the absolute disparity test, which measures the magnitude of underrepresentation by the difference between the percentage of a cognizable group in the venire, and the percentage of the group in the general population. A *cognizable group* consists of persons who self-identify as a group, and can be recognized by outsiders as members of the group. Gender, race, and ethnicity form cognizable groups; religious orientation *might* form the basis of such a group, particularly for religions like Orthodox Judaism or Sikhism that mandate distinctive dress. Sexual orientation might also qualify as the basis for a cognizable group. In *Swain*, a majority of the justices held that even so large a disparity as 10% did not constitute prima facie evidence of discrimination.

The absolute disparity test became far more important following the Supreme Court's ruling in *Duren v. Missouri* (1979). The petitioner, Duren, had been convicted of murder in Jackson County, Missouri. While 54% of the residents of the county were women, only 26.7% of those who received summonses were women; and only 14.5% of those who actually appeared for jury duty were women; Duren's 53-person venire included five women, none of whom served on the actual jury.

This peculiar situation arose because Missouri permitted women to exclude themselves from jury service; moreover, women who failed to show up in response to a summons were automatically excused. Ruth Bader Ginsburg, who would become an Associate Justice of the Supreme Court in 1993, argued the case before the Supreme Court, charging that the underrepresentation of women on the venire violated the fair cross-section requirement, and thus infringed on Duren's 6th Amendment rights. The Supreme Court agreed in an 8 to 1 opinion (Justice Rehnquist dissenting). In the majority opinion, Justice Byron White established what is now known as the *three-pronged Duren test*.

In order to establish a prima facie violation of the fair cross-section requirement, the petitioner must show that the group excluded is a "distinctive" group (i.e., it is a cognizable group); that there is a substantial disparity between the representation of this group in the venire and the population;

and that this disparity is caused by a systematic exclusion of the group in the jury selection process. Put together, *Duren* and *Swain* have been interpreted to mean that a small disparity between the percentage of a cognizable group in jury venires and the percentage of the group in the general population will not constitute a violation of the fair cross-section requirement. The Supreme Court has repeatedly denied having set any specific percentage. However, other courts have seized on the "ten percent underrepresentation" referred to in *Swain* as a *de minimis* (minimal) requirement.

There are several objectionable features of absolute disparity. One obvious problem is that it sets an "allowable" level of discrimination that would be dismissed by a higher court (barring open admission of discriminatory selection of jurors). One could easily imagine jury commissioners drawing up venires carefully, so as to underrepresent a group by less than the 10% threshold. In fact, one need not imagine this: the appeal in *Amadeo v. Zant* (1988) centered around a memorandum that suggested that the district attorney for Putnam County, Georgia did precisely that, giving jury commissioners quotas for the numbers of women and blacks to place on the venires so that these groups would be underrepresented on jury venires by between 5% and 11%— just within the limits mentioned in *Swain*.

The magnitude of this "allowable" discrimination can be illustrated as follows. Suppose, as in *Swain*, that blacks constitute about 25% of the eligible jurors in a county, and the following procedure is used to draw up a jury venire. First, a preliminary venire is drawn by randomly choosing names from a list of eligible jurors. Next, the list is examined, and one-third of the blacks are removed from the list. Under this system, blacks would make up just over 18% of the resulting venire, so "We cannot say that purposeful discrimination based on race alone is satisfactorily proved"!

The Flaw of Large Numbers

Supporters of absolute disparity point out that it is easy to use, requiring only grade school arithmetic. But this argument is specious: If we expect judges and jurors to be adults, why limit them to what they learned as children?

A more sophisticated approach would use statistical decision theory. However, this presents a different problem: do we focus on the underrepresentation for the venire that tries a specific defendant (as in *Avery*), or do we focus on the underrepresentation for all venires (as in *Swain*)? Either approach leads us to difficulties.

To understand these difficulties, consider the problem of determining whether a coin is fair or not. Rather than perform a set number of tosses, let us instead flip the coin until we suspect the coin is unfair, or until the evidence suggests otherwise. This is analogous to how the law makes the same sort of decision regarding jury discrimination.

If we only flip the coin once, we will have a substantial disparity between what we expect (50% heads) and what we observe (0% heads, or 100% heads, depending on how the coin lands). If we flip the coin twice and it comes up the same way both times, we might suspect the coin is unfair; however, most people would probably agree that there is as yet insufficient evidence for unfairness. If we flip the coin three times and it comes up the same way all three times, there is somewhat stronger evidence that the coin is unfair, and so on. The question to consider is: how many consecutive heads (or tails) must we see before we can reasonably conclude the coin is not fair?

The answer, of course, depends on the false alarm rate we are willing to accept. The highest commonly accepted false alarm rate is 5%, but the decision in *Vasquez* suggests that 0.3% is a more palatable standard for the law. If we insisted on a false alarm rate of 0.3% or less, then nine consecutive heads (or nine consecutive tails) will not be enough evidence to conclude the coin is unfair! Put another way: if we only flip the coin nine times, *no* disparity will be enough to reject the null hypothesis.

This was the situation in *Avery*, where a venire of 60 was drawn from a population that was 5% black. In this case, even total exclusion of blacks from the venire would not be regarded as statistically significant; Justice Frankfurter's well-meaning comment aside, a purely random drawing of the veniremen could easily produce an all-white venire. Indeed, if we hold to the 0.3% false alarm rate suggested by *Vasquez*, we would find that even total exclusion of blacks from a 150-person venire drawn from a population that was 5% black could reasonably be attributed to chance. Thus a statistical approach may make it very difficult to establish a prima facie case of discrimination on the basis of just one venire.

What if we look at all venires? Here we confront a different problem. Suppose we give a large number of people fair coins, and have them flip the coin 1,000 times. Most of our experimenters would see the coin land heads up between 450 and 550 times. But a disparity as large as 10%—say seeing fewer than 400 heads in 1,000 flips—would be seen by one experimenter in perhaps four billion.

While this suggests that the probability of observing a sizable disparity is very low, it also means that almost any difference between the observed and expected value will be regarded as statistically significant. This may result in an impossibly high standard for the state to meet when selecting jury venires since, although the state can issue summonses for jury service, it can do little to ensure that those summoned will actually appear. Even a slight difference in the appearance rate can result in a statistically significant disparity.

For example, suppose 10% of the eligible population is black. If over some interval of time the state sends summonses to 10,000 people, and all of the whites appear but only 90% of the blacks do, we would observe a venire of 9,900 that included 900 blacks. The number of blacks we observe on the venire is slightly more than three standard deviations below the number we would expect to observe, and a statistician would conclude the disparity is statistically significant. To make matters worse, the state cannot easily defend itself, because persons who register for jury service are not required to indicate their race. Even if it were true that persons of one race showed up less frequently than persons of another, there would be no way to show this using the available evidence.

Comparative Disparity

Another problem with absolute disparity occurs when a cognizable group makes up less than 10% of the population. One might hope that common sense would assert itself and reject the application of absolute disparity in such cases. One would be disappointed: the Seventh Circuit Court of Appeals rejected a discrimination claim in *U.S. v. Phillips et al.* (2001), in part because the disparity did not exceed 10%—even though blacks made up only 6% of the eligible population.

Judge Nancy Gertner, of the U.S. District Court for the District of Massachusetts, expressed her dissatisfaction with absolute disparity in *U.S. v. Green* (2005) and urged the 1st Circuit to consider supplementing absolute disparity with other statistical measures. She considered three in particular. The first appeared in a 1973 Civil Rights Commission report, which suggested that a disparity of more than 20% between the proportion of eligible white veniremen and the proportion of eligible minority veniremen should be remedied by supplementation (i.e., deliberate recruitment of more minority members). For example, if 70% of the white population was eligible, jury commissioners should act to ensure that at least 56% of the minority population was also eligible.

Supplementation is based on *comparative disparity*, which focuses on the relative difference between the observed and expected number of jurors belonging to a group, using the expected number of jurors belonging to the group as the reference quantity. Since the report recommended supplementation for disparities that exceeded 20%, we might take a comparative disparity that exceeds 20% as prima facie evidence of discrimination.

Comparative disparity appears to be useful when a cognizable group forms a small fraction of the population. However, the Court of Appeals for the 1st Circuit (Massachusetts, New Hampshire, and Maine) presented a hypothetical example that points to a problem with comparative disparity. In *U.S. v. Hafen* (1984), the Court considered a hypothetical situation where a jurisdiction had 500,000 whites and 1 black. Failure to include the single black person on the venire would result in a 100% comparative disparity, even though an all-white jury would, by most reasonable standards, be a "fair cross section" of the community. This decision effectively prohibited Gertner from using comparative disparity in *U.S. v. Green*.

However, the *Hafen* decision, while it set precedent, only bound the *federal* courts of the 1st district. State courts, even those geographically part of the 1st district, could accept or reject comparative disparity, and while their decisions could be reviewed, they would not be a priori impermissible. Moreover, federal courts in other districts could also choose their own standards. Again, only a decision by the Supreme Court would be binding on all courts in the United States.

Hence the importance of *Berghuis v. Smith*. Diapolis Smith would be tried in Kent County, Michigan, where 7.28% of the jury eligible population was black. In the six months before the trial, 929 veniremen had been drawn, but only 56 (about 6%) were black. Smith's own venire of between 60 and 100 individuals (the court records are incomplete) had only 3 blacks, none of whom would be empaneled.

Obviously, since blacks made up only 7.28% of the population, it would be impossible to show an absolute disparity of more than 10%. Here there was an absolute disparity of 1.25% (for the months leading up to the trial) and 2.28%

(for the trial itself), which, by the 7th Circuit's reasoning in *Phillips*, could be disregarded. Moreover, since there were some blacks on both the preceding venires and the venire that tried Smith, total exclusion did not occur, much like the situation in *Avery* and *Williams*. Thus the Michigan State Supreme Court ruled against Smith's claim that the jury had been improperly selected.

However, the comparative disparities were 18% (for the months leading up to the trial) and 31% (for the trial itself). Put another way, roughly one-sixth of the blacks that should have been on the venires in the preceding months, and roughly one-third of the blacks that should have been on the trial venire, were absent. When the 6th Circuit Court of Appeals heard Smith's appeal, it accepted an argument based on comparative disparity, holding that Smith's right to a jury drawn from a fair cross-section of the population had been violated. The court reversed and remanded.

This was a remarkable decision, for the *Hafen* example has caused many jurisdictions to reject the use of comparative disparity as a means of evaluating underrepresentation. The state of Michigan appealed to the Supreme Court; it went further and called on the Supreme Court to adopt a general requirement of a 10% absolute disparity in all future fair cross-section cases.

The Supreme Court heard the case on January 20, 2010. In a unanimous decision, it overturned the Court of Appeals, effectively upholding Smith's conviction. However, the Supreme Court declined to endorse absolute disparity; in fact, it refused to endorse any particular method of evaluating underrepresentation.

The Eighty Percent Rule

The Supreme Court's decision in this case is particularly baffling, because it had *already* endorsed a method of evaluating underrepresentation in another context: employment discrimination. In *Griggs v. Duke Power Company* (1971), the Supreme Court ruled that policies that reduced the eligibility of blacks for promotion had to be demonstrably related to job performance.

In *Hazelwood v. United States* (1977), the Supreme Court went further. As in jury discrimination cases, we can expect some variation in the selection rates; thus some level of underrepresentation might be a result of purely random factors. *Hazelwood* concerned the hiring rates of blacks in the Hazelwood School District (just outside of St. Louis, Missouri), and the Court ruled that statistics alone could establish a prima facie case of discrimination.

The *Griggs* and *Hazelwood* rulings would be formalized by the Equal Employment Opportunity Commission (EEOC) as the "80% rule" (also known as the four-fifths rule): A prima facie case of *adverse impact* can be established if a policy selects a cognizable group at a rate less than 80% of the selection rate of the most favored group, *and* this lower selection rate is statistically significant. Once the prima facie case of adverse impact is established, the burden of proof shifts to the employer, to show that the policy satisfies a business necessity (in effect, that the requirements are job-related).

The 80% rule is clearly related to the 1973 Civil Rights Commission recommendation for remedying minority underrepresentation on venires, and can be applied easily enough. For example, many states require literacy in English as a requirement for jury service. This would tend to select recent immigrants from non–English-speaking countries at a lower rate than the general population, and the difference would likely be statistically significant; thus one could easily establish a prima facie case that the requirement had an adverse impact on these groups. The burden of proof would shift to the state, to show that the requirement is performance related. In this case, a literacy requirement for jurors is a reasonable one, so while this requirement would produce underrepresentation of some minority groups, it would probably be permitted.

In contrast, consider the requirement in *Patton* that jurors be registered voters; at the time, this meant presenting proof that you had paid your poll tax. In this case, it is not as clear that performance as a juror has anything to do with being able to prove that you paid a poll tax.

To see the flexibility of the 80% rule, consider the hypothetical situation in *Hafen*, where a jurisdiction had 1 black and 500,000 whites. Consider a venire of 100 with no blacks. Since blacks constitute $1/500{,}001 \approx 0.002\%$ of the population, their absence from the venire amounts to a 0.002% absolute disparity— far below the 10% suggested in *Swain*. We would not, on the basis of absolute disparity, be able to press a claim of jury discrimination.

On the other hand, since about 0.002% of the population is black but 0% of the veniremen are black, we have a comparative disparity of 100%. If we grant that *some* level of comparative disparity can be used to establish a prima facie case of a fair cross-section violation, and since the total exclusion of a cognizable group would give us the highest possible comparative disparity, it follows that this situation would automatically constitute a prima facie case of a fair cross-section violation.

However, as the 1st Circuit Court of Appeals pointed out, an all-white venire seems to represent a fair cross section of the community. Comparative disparity is thus in the position of the boy who cried wolf: it loses credibility every time it draws attention to situations that no reasonable person would find objectionable.

A sizable comparative disparity is therefore not enough. The second part of the 80% rule requires us to assess whether the underselection is statistically significant. In this case, we find that it is not, so we would conclude that there is insufficient evidence to conclude there has been a fair cross-section violation.

The 80% rule allows us to go further. Both comparative disparity and absolute disparity avoid one important issue: defendants are tried by different juries, selected from different venires. In *Avery*, blacks *did* appear on some venires—just not the venire that tried *Avery*. Thus we must confront the question of which figure is relevant: underrepresentation of a cognizable group in the general venire, or underrepresentation of a cognizable group in the venire for a specific trial.

There are two ways that a cognizable group's presence on a jury can be reduced or even eliminated. First, they might be underselected for the general venire: if there are no black veniremen, there can be no black jurors. This was the situation in *Patton*. Next, the general venire might be chosen without regard to race, but a cognizable group might be underselected for a particular venire. This was the situation in *Avery* and *Smith*.

Strauder established the principle that only the first is important. But this leaves open the possibility that the prosecutor might "stack the deck" to ensure a particular verdict in a high-profile case. For example, consider the case of Diapolis Smith. In Kent County, Michigan, where the case was tried, 7.28% of the jury-eligible population was black, but during the six-month period leading up to the trial, 56 of the 929 veniremen were black. Thus blacks were selected at 82% the rate of whites.[3] Since this exceeds the 80% selection rate of whites, it would not be possible (under the 80% rule) to establish a prima facie case of discrimination on the basis of the selection rate alone.

---

3. We can determine the selection rate of a group as follows: If the venire consists of r members of a cognizable group and s members of the most favored group, while the percentages of these groups in the general population are p and q respectively, the selection rate is $k = (r/s) / (p/q)$.

What about Smith's own venire? The actual size of the venire was not re-corded, but it was believed to have been between 60 and 100, and between two and three were black. If 100 individuals were called, only three of whom were black, the selection rate of blacks would only be 40% of the selection rate of whites. Even if only 60 veniremen were called, again with three blacks, the selection rate of blacks would only be 67% of the selection rate of whites.

Since this falls below the 80% selection rate of whites, the second part of the 80% rule would be invoked, and we would ask whether the disparity was statistically significant. In this case, the number of blacks does not exceed the "two or three" standard deviations that would ordinarily be taken as statisti-cally significant. We might conclude that the disparity was not statistically significant.

Disparity of Risk

The 80% rule is already accepted by the EEOC as a means of establishing a prima facie case of adverse impact. Absolute disparity is generally accepted by the courts, and the Supreme Court has at least not rejected comparative disparity. However, there are other proposals to measure jury discrimination which have yet to be endorsed by any court. One of the more promising is known as the disparity of risk, proposed by Peter A. Detre, PhD, a mathema-tician turned lawyer, and referenced by Gertner in her decision in *U.S. v. Green*.

Disparity of risk is based on the idea that, if a cognizable group is under-represented in the venires, there is an increased risk that the group will be underrepresented on the jury panel itself. To determine this increased risk, we can calculate two sets of probabilities: first, the probability that the jury will have n or fewer members of the identifiable group on it, if drawn from an "ideal venire" containing the expected number of members of the distinctive group (we compute this for all values of n); second, the same probability for a jury drawn from the actual venire. Finally, we can find the difference be-tween the two probabilities.

For example, consider the *Smith* case, where the venire was drawn from a population that was 7.28% black. Assume the venire had 60 persons, 3 of whom were black. If we randomly draw 12 persons from this venire to form our jury, we can determine the probability that the jury contains 0, 1, 2, 3, 4, etc., blacks: these probabilities will be 50.54%, 39.56%, 9.26%, and 0% respectively (the last because a jury containing four or more blacks cannot be drawn from a venire containing only three). However, an ideal venire that reflected the composition of the population would be 7% black; hence a venire of 60 would include 4 blacks. If we randomly drew 12 persons from this ideal venire to form our jury, the probability that it contains 0, 1, 2, 3, or 4 blacks would be 39.90%, 42.56%, 15.27%, 2.17%, and 0.10%.

At this point, we confront the problem of evaluating the significance of the probabilities. Detre suggests using the greatest difference in probabilities as a measure of the risk. In this case, the actual venire made the defendant (Smith) about 11% more likely to face a jury with no blacks. However, it is not clear how to translate this increase in risk into a legally meaningful standard.

Even more important, as case after case shows, merely including minority members in the venire will not guarantee their presence on the jury itself. As a result, regardless of whether the venire is a fair cross section of the population, opportunities exist to reduce or even eliminate minority presence on the jury panel. We address this issue in the next section.

Once Is an Accident . . .

> Given an unencumbered right to exercise peremptory challenges, one
> might expect each party to attempt to eliminate members of those
> groups which are predisposed toward the opposition. However, when
> the defendant is a minority member, his attempt is doomed to failure.
> —PAUL LIACOS, ASSOCIATE JUSTICE, *COMMONWEALTH*
> *OF MASSACHUSETTS V. SOARES* (1978)

The U.S. Supreme Court's ruling in *Swain* actually covered two issues regarding petit juries. We have discussed the flaws of the ten percent rule in detail. But *Swain* also ruled on another issue: peremptory challenges, where either side may eliminate ("strike") a potential juror without having to give a reason. Obviously, a prosecutor who wishes to discriminate against members of a cognizable group could use the peremptory challenges to do so. In fact, most states give both sides generous numbers of peremptory challenges in criminal cases, which makes it even easier to deliberately exclude all members of a cognizable group from a jury panel.

In *Swain*, after excusals and dismissals for cause, 75 veniremen remained, including six blacks. At this point, both prosecution and defense began to strike jurors, with the defense allowed to strike twice as many jurors as the prosecution. The prosecutor used six (of his 21) peremptory challenges to eliminate all blacks from the venire, resulting in an all-white jury. It takes no great imagination to suppose the jurors were struck because they were black. But the Supreme Court held that, because neither side needed to give a reason for issuing a peremptory challenge, the prosecution need not defend their usage in any particular instance.

Fortunately, the states could set higher requirements than the Supreme Court, so California became the first state to grapple with an important civil rights issue. In *People v. Wheeler* (1978), the Supreme Court of California considered the case of James Michael Wheeler, convicted of murdering Amaury Cedeno during a robbery. The state charged that Wheeler had driven the getaway vehicle. The case against Wheeler was weak at best: the only evidence linking him to the getaway car consisted of two fingerprints on the driver's side door (with no indication of when the fingerprints had been placed).

Although at least seven blacks were on the venire drawn for Wheeler's case, the prosecution used his peremptory challenges to eliminate them.[1] Thus Wheeler, a black man charged with murdering a white man, would be convicted by an all-white jury. The Supreme Court of California held that while it was true that no reason needed to be *given* for a peremptory challenge, it did not follow that no reason *existed*. And if the reason not given was the race of the juror challenged, then the defendant's rights to a fair trial (granted by the constitution of the state of California) were being violated. The court overturned Wheeler's conviction, and the state declined to re-prosecute.

In *Commonwealth v. Soares* (1979), the Supreme Judicial Court of Massachusetts considered a similar case. In the early morning hours of November 16, 1976, a group of Harvard College football players, leaving a strip club in Boston's "Combat Zone" (at the time, one of the less savory neighborhoods of downtown), got into an altercation with three black men. The football players had been chasing a black woman, accusing her of having stolen a wallet from one of them; the men intervened, and in the ensuing fight, Andrew Puopolo, one of the football players, was stabbed in the heart; he died a month later. The black men, Leon Easterling, Edward Soares, and Richard Allen, were indicted for murder in the first degree, convicted on all counts, and given mandatory life sentences.

---

1. Ironically, to avoid the possibility of rejecting jurors on the basis of race, venirepersons are not required to announce their race, religion, or ethnic origin during voir dire, nor is this information otherwise recorded. As a result, the exact number of blacks summoned and dismissed in the Wheeler case is unknown; only afterward did the defense establish that at least seven of those dismissed were black.

The venire contained 13 blacks and 94 whites. However, the prosecutor used 12 of his 44 peremptory challenges to exclude all but one black from the final jury panel. The defense argued that the blacks were eliminated because the prosecutor believed that black jurors would be more sympathetic to a black defendant. In a landmark decision, the Supreme Judicial Court of Massachusetts ruled: "exercise of peremptory challenges to exclude members of discrete groups solely on the basis of bias presumed to derive from that individual's membership in the group, contravenes the requirement inherent in article 12 of the Declaration of Rights" (377 Mass. 461, p. 488). Consequently it overturned the conviction. Upon selection of a new venire and a new jury panel, Easterling was found guilty of manslaughter and was given an 18–20-year sentence.[2] The other co-defendants were acquitted.

The Supreme Court came around to this point of view in *Batson v. Kentucky* (1986). In this case, the venire consisted of 28 persons, including 4 blacks; the prosecutor had six peremptory challenges and the defense nine, which were used to reduce the venire to a jury panel of twelve, plus one alternate. The prosecutor used four of his six challenges to eliminate all blacks from the panel. In a 7 to 2 decision, the Supreme Court ruled that if there was prima facie evidence that the peremptory strikes were made on the basis of race, the burden of proof shifted to the prosecutor to provide race-neutral explanations for the strikes.

Fisher's Exact Test

Batson established the principle that the selection of jurors for a specific trial might be sufficient to establish a prima facie case of jury discrimination. But, as with discriminatory selection of jurors for the venire, we must ask: how much disparity is required before we have prima facie evidence of race-based discrimination that must be rebutted by the state? A key problem is that in general, the number of allowed peremptory strikes is small, and it is a truism that statistics only works for large samples.

As with many truisms, this is an oversimplification: statistics *can* be applied to small samples. One way to handle small samples is to use *Fisher's Exact*

---

2. Easterling escaped in October 1986 during a work-release job, a controversial program that dogged the 1988 presidential campaign of Massachusetts governor Michael Dukakis.

Table 12.   Prosecutorial Peremptory Challenges from *Swain*

	Black	Non-Black	Total
Dismissed	6	15	21
Not Dismissed	0	54	53
Totals	6	69	75

Test, which is based on computing hypergeometric probabilities from a *2 × 2 contingency table.*

As the name suggests, a 2 × 2 contingency table organizes the data into a table with two rows, based on whether a given factor is present or not; and two columns, based on whether another factor is present or not. In this case, the two factors would be the race of the juror and whether or not they were dismissed; the *Swain* data are shown in table 12.

In *Swain,* the prosecutor struck 21 persons from a venire of 75 that included 6 blacks. If the strikes were done at random, the probability of eliminating all 6 blacks would be 0.0002695, or about 1 in 4000. As before, if we view this event as too improbable to occur by chance, we ought to consider other events with equal or lower probabilities. Thus we should compute the probability that 5, 4, 3, 2, 1, or 0 blacks are dismissed peremptorily. As it happens, all of these are more probable than the observed event.

What we actually observed, which was six blacks out of six challenged peremptorily, had a probability of 0.0002695. If we reject the null hypothesis, we set a precedent for rejecting the null hypothesis any time we observe an event with probability 0.0002695 or less; however, there are in this case no events with a lower probability. We can conclude that the probability of making a Type I (false alarm) error is less than 0.0002695. This is low enough that we would say the results *are* statistically significant and that choosing who to strike and who to retain is *not* done randomly. However, the Supreme Court upheld the sanctity of the peremptory strike and ignored mathematics entirely to produce a decision that we must regard as statistically suspect.

In *Batson,* the venire of 28 included 4 blacks, and the prosecutor used four of his six challenges to strike all of them. The probability of such an extreme result is about 0.07%, so such a pattern would occur about once in 1,300 similar cases. The probability of observing a pattern of strikes like that in *Soares* is even lower: about one in 14,000. In all three cases, there is sufficient evidence to reject the null hypothesis that the strikes were done at random.

There are two objections to the preceding analysis. The first is that rejecting the null hypothesis *only* entails accepting that the strikes were not done at random; it does *not* mean that the strikes were done on the basis of race. The second is that Fisher's Exact Test is based on selecting a set of objects from a larger set of objects, where the selection is done "all at once." It can be (and has been) applied successfully in employment discrimination cases, where candidates are selected from a group of applicants. But in *Swain*, peremptory challenges were issued alternately by the prosecutor and defense attorney, and in most jurisdictions, the challenges can be issued by either side at any point in the course of voir dire.

In 2007, Bruce Barrett of the University of Alabama at Tuscaloosa, suggested that the *Poisson binomial distribution* be the basis of the statistical analysis of peremptory challenges. The details of using the Poisson binomial distribution are somewhat complex, but the basic problem may be illustrated as follows. Imagine a defense attorney who always strikes male jurors and a prosecutor who issues strikes at random without regard to gender. Suppose our venire consists of equal numbers of men and women. Every time the defense attorney strikes a prospective male juror through a peremptory strike, it leaves the prosecutor with a venire that has slightly more women than men—which means that the prosecutor, who is issuing strikes at random without regard to gender, is more likely to strike a female venireperson. In other words, the discriminatory behavior of the *defense attorney* can create an appearance of discriminatory behavior by the *prosecutor*.

Thus, Fisher's Exact Test should be used only when one side makes it selections without interference from the other. For example, if the prosecution and defense wrote down the names of all the jurors they would strike, without knowledge of which jurors would be struck by the other (and thus leaving open the possibility that some jurors would be struck by both sides), then Fisher's Exact Test could be applied.

One situation where it might have been applicable was in *Miller-el v. Cockrell* (2003). Thomas Joe Miller-el had been convicted of murdering Doug Walker, a Holiday Inn employee, during a robbery in Dallas, Texas; Miller-el was convicted and sentenced to death. His appeal centered on the prosecution's use of peremptory challenges. After excusals and dismissals, the venire was reduced to 42 persons, including 11 blacks. The prosecution had 14 peremptory challenges, and used 10 of them to strike all but one of the black veniremen. Nevertheless, the Supreme Court of Texas held that Miller-el presented insufficient evidence to conclude that the peremptory challenges were based on race, and upheld the conviction and death sentence.

Following *Batson*, the prosecutor gave race-neutral explanations for the strikes. Some of these were issued on the basis of whether a venireman expressed ambivalence or opposition to the death penalty. Since a guilty verdict would put Miller-el at risk of being sentenced to death, one could argue that such ambivalence might affect a juror's objectivity and, following the logic of employment discrimination cases, be a reasonable basis for rejecting jurors.

However, there were some anomalies in the conduct of the voir dire phase. Some jurors were presented with an abstract scenario that simply indicated the state was seeking the death penalty, while others were given a much more graphic description of how the penalty would be carried out. The more graphic script was presented to 3 of the 31 white veniremen (about 9%), and 6 of the 11 black veniremen (about 53%).

This is a situation where Fisher's Exact Test is applicable, since the decision of who will be presented with the graphic or the abstract scenario can be treated as selecting a set of persons from a larger set of persons "all at once." Here we find that if veniremen were randomly assigned one of the scenarios, there is only a 1 in 200 chance of producing such a large disparity between the fraction of blacks who received the graphic script and the fraction of whites who did so. Most statisticians would conclude that some non-random factor was in operation.

This need not mean that race was the factor that decided who would receive the graphic script and who would receive the abstract script. The prosecution argued that the decision was based on jurors who expressed ambivalence about the death penalty on a pretrial questionnaire. Ten non-blacks (32%) and seven blacks (64%) expressed ambivalence about the death penalty in a pretrial questionnaire. But in the final analysis, not all jurors who expressed ambivalence received the more graphic script. Six of the seven blacks (86%) but only three of the ten non-blacks (30%) received the more graphic script.

The Supreme Court seized upon this disparity as the important one. Justice Souter, writing the majority opinion, noted that "if we posit instead that the prosecutors' first object was to use the graphic script to make a case for excluding black panel members opposed to or ambivalent about the death penalty, there is a much tighter fit of fact and explanation" (545 U.S. 231, p. 260).

The Supreme Court remanded the case, and a plea-bargain agreement ended with Miller-el serving a life sentence for murder and a 20-year sentence for aggravated robbery, to be served consecutively.

Souter's remark and the Supreme Court's decision raise an important philosophical question. As before, if we rule that the observed event is too improbable to occur by chance, we set the precedent that any event less likely to occur is also too improbable to occur by chance. It turns out that the observed event (6 of 7 blacks, and 3 of 10 non-blacks, receiving the graphic script) has a probability of 0.03455. Given the same group of jurors and the same numbers assigned to receive the graphic and the abstract scripts, we find three events with a lower probability:

- The probability that 7 of 7 blacks, and 2 of 10 non-blacks, received the graphic script: 0.004936.
- The probability that 1 of 7 blacks, and 8 of 10 non-blacks, received the graphic script: 0.004936.
- The probability that 0 of 7 blacks, and 9 of 10 non-blacks, received the graphic script: 0.004936.

Thus rejecting the null hypothesis in Miller-el's case corresponds to accepting a false alarm rate as high as $0.03455 + 0.004936 + 0.03455 + 0.004936 = 0.07898$ (7.898%).

Note that this is higher than 5%, the highest commonly used false alarm rate. A statistician might conclude that the differential treatment of the ambivalent jurors was not statistically significant: in other words, it could have been produced by chance alone. The Supreme Court, on the basis of such testimony, might have dismissed the disparity and allowed Miller-el's death sentence to stand.

Does this mean that the Supreme Court's decision in *Miller-el v. Dretke* was statistically flawed? Not at all. Remember that our choice of a false alarm rate depends on the consequences of a false alarm versus the consequences of a missed alarm. In the employment discrimination case *Kadas v. MCI* (2001), Justice Posner correctly noted:

> The 5 percent test is arbitrary; it is influenced by the fact that scholarly publishers have limited space and don't want to clog up their journals and books with statistical findings that have a substantial probability of being a product of chance rather than of some interesting underlying relation between the

variables of concern. Litigation generally is not fussy about evidence; much eyewitness and other nonquantitative evidence is subject to significant possibility of error, yet no effort is made to exclude it if it doesn't satisfy some counterpart to the 5 percent significance test. (255 F.3d 359, p. 362)

In *Miller-el v. Dretke*, the consequences of a false alarm would be a retrial of Miller-el; in fact, rather than face retrial, Miller-el agreed to a plea bargain and will spend the rest of his life in prison for robbery and murder. In contrast, the consequences of a missed alarm would have been the execution of Miller-el for robbery and murder after conviction by a biased jury. In this situation, we should regard a missed alarm as much worse than a false alarm. We should try to obtain as low a missed alarm rate as possible. This translates into accepting a high false alarm rate: thus it would be reasonable to use a false alarm rate of 7.898% or even higher.

Engineering the Peremptory Challenge

Although statistics can be used to establish a prima facie case of race-based peremptory challenges, the state can still attempt to rebut the charge by presenting race-neutral reasons for juror dismissal. For example, consider *Lockett v. Mississippi* (1987), heard by the Supreme Court of Mississippi. Carl Daniel Lockett (who renamed himself Shaka Daniel Muhammad al-Zulu following his conversion to Islam in 1992) had been convicted of murdering John and Geraldine Calhoun in 1986. During voir dire, the prosecutor used five of his first seven peremptory challenges to strike all blacks from the venire.

Following *Batson*, the prosecutor was called on to provide race-neutral explanations for the strikes, and he obliged. One of the black men was a preacher; one woman's brother was a convicted felon; yet another expressed concerns about sequestration. The prosecutor dismissed a single 22-year-old laborer with an eleventh grade education because his youth, marital status, and educational level suggested instability. The fifth black was struck because he wore a hat into the courtroom, which suggested a general disrespect for the proceedings. The Mississippi Supreme Court accepted these explanations and upheld the conviction.

But if "once is an accident, twice is a coincidence, three times is a trend," we might suspect that five race-neutral explanations that happened to apply to the five blacks on the venire might be something more. Judge James L.

Robertson, in his dissent, noted that Fisher's Exact Test gave a "level of significance of 0.00000000397." Robertson probably meant that if we used a level of significance of 0.00000000397, we would reject the null hypothesis and conclude that some non-random factor led to the striking of blacks from the venire. Hence, Robertson argued, even if the *individual* strikes could be explained using race-neutral explanations, the fact that the race-neutral explanations "just happened" to apply to the minority members is suspicious.

One possible solution is to reduce the number of peremptory challenges or eliminate them entirely, which Justice Marshall suggested in *Batson*. Peremptory challenges are supported by a thousand years of Anglo-Saxon jurisprudence, but *not* by the Constitution, so abolishing them is not inconceivable. Since the number of peremptory challenges is set by state and federal law, it would be easy enough to reduce or even eliminate them.

However, doing so may be counterproductive, in that it might make it impossible to establish a prima facie case of discrimination using statistics alone. For example, suppose the prosecution and defense each had one peremptory challenge. Then a jury of 12 would require the summoning of a 14-person venire (actually more, since we must assume some jurors will be dismissed or excused for cause). If the venire contained one member of a minority group, there would be about a 7% chance that the minority member would be struck by a purely random application of the peremptory challenges by the prosecution. Consequently, such an event could easily be attributed to chance. Moreover, as *Lockett* shows, prosecutors called upon to produce a race-neutral explanation for *one* strike can easily do so.

In contrast, suppose the venire was ten times large with ten times as many minority members. To reduce the venire to a trial jury, both sides would have 64 peremptory challenges. In this case, if the prosecution used the strikes completely at random, there is about one chance in 4,000 that all ten minority members would be eliminated. If all ten *were* eliminated, this would be statistically significant evidence that the strikes were not random with regard to race. Moreover, the prosecution would have to produce several race-neutral explanations for the strikes. While *Lockett* shows that this could be done, Robertson's dissent shows that at some point, the pattern of strikes itself becomes suspicious.

Rather than abolishing peremptory challenges altogether, we might instead engineer the number so that using them in a race-based fashion is counterproductive. Even before *Batson*, statistician Joseph Kadane of Carnegie-

Mellon and lawyer David Kairys of Philadelphia considered the problem and suggested the following solution. It is generally held that peremptory challenges exist to give both sides the opportunity to eliminate jurors that might be biased, based on nothing more than intuition that a particular venireperson might not judge a case objectively. Presumably, the prosecution would not intentionally strike venirepersons biased in favor of the state, nor would the defense intentionally strike venirepersons biased in favor of the defendant.

Suppose each side has just enough peremptory challenges to eliminate venirepersons biased against that side. Then neither would waste a challenge for spurious reasons. If a prosecutor struck venirepersons because of their race, he or she would not have enough peremptory challenges to eliminate all venirepersons biased against the state. Thus the final jury panel would not be objective—it would be biased in favor of the defendant (who would, of course, be under no obligation to challenge an acquittal!).

To determine the optimal number of peremptory challenges, let r be some threshold probability, and let prosecution and defense have x and y peremptory strikes, respectively. In addition to the jurors (and alternates) who will actually sit in judgment, the venire must also contain an additional $x + y$ persons after excusals and dismissals for cause. We want to find x, y so that the probability that there are x or fewer jurors biased against the prosecution, and y or fewer jurors biased against the defense, is at least r. Both sides will have probability r (at least) of being able to strike all jurors biased against them. At the same time, if either of them uses a peremptory challenge for any other reason, they will have a substantial probability of facing a jury that is biased against them.

Setting r is purely a matter of legal philosophy. Kadane and Kairys suggest using 95% for noncapital criminal cases, and 99% for cases that might result in the death penalty. However, determining x and y from the chosen value of r requires us to know the probability that a venireperson is biased against one side or the other.

For example, suppose 10% of the prospective venirepersons, after excusals and dismissals for cause, are biased against the prosecution, and a similar fraction biased against the defendant (with the remaining 80% of the venirepersons truly objective). If each side has one peremptory challenge, then after excusals and dismissals for cause, there would have to be $12 + 1 + 1 = 14$ venirepersons left. Under these circumstances, the probability that one or fewer persons is biased against the prosecution is about 58%. This means there is a 42% chance that there will be more than one venireperson biased against the prosecution. Because the prosecution only has one peremptory challenge, this means that 42% of the time, it will not be able to eliminate all jurors biased against it. The defense confronts the same problem.

On the other hand, if both sides had four peremptory challenges, a venire of $12 + 4 + 4 = 20$ would have to remain after excusals and dismissals for cause. In this case, there is a 95% chance that the remaining 20 members of the venire will include four or fewer persons biased against the prosecution, and so a 95% chance that the prosecution will have enough peremptory challenges to strike all venirepersons biased against it. The defense has a similar opportunity.

Unfortunately, the fraction of the venire biased toward one party or the other seems impossible to determine. As an alternative, we might begin by supposing that the number of peremptory challenges *will* in general produce an impartial jury from a venire drawn from the population at large, then increasing the number for more serious (e.g., capital) cases.

As it turns out, most states do increase the number of peremptory challenges for capital cases (Kentucky, along with Virginia and South Carolina, are notable exceptions). However, the increased number seems to be based more on legal intuition than on mathematical analysis.

For example, the state of Mississippi permits both sides 6 peremptory challenges in noncapital felony cases and 12 in capital cases. If we assume that the 6 peremptory challenges in felony cases were meant to give both sides a 95% chance of striking all venirepersons biased against it, this translates into a 14.57% chance that a venireperson is biased against either side.

In a capital case, $12 + 12 + 12 = 36$ venirepersons would have to remain after excusals and dismissals for cause. If the probability that a venireperson is biased against one side is 14.57%, then there is a 99.89% chance that both sides will be able to eliminate all venirepersons biased against it. Thus, the increase in the number of peremptory challenges seems to be consistent with the needs of justice.

There is a danger inherent in this system. Suppose the prosecutor struck two jurors because of their race. This would leave 10 challenges to eliminate any potential jurors biased against the prosecution. But there is a 98.89% chance that the venire contains 10 or fewer jurors biased against the prosecution. Thus the prosecutor could well afford to waste peremptory challenges for spurious reasons.

But what if we reduce the number of peremptory challenges? For example, if both sides had nine peremptory challenges, they would have a greater-than-99% probability of being able to strike all jurors biased against it. But if the prosecution used even a single strike to eliminate a juror on the basis of race, he or she would run a risk (about 2.3%) of not having enough strikes to eliminate all jurors biased against it. In this way, we engineer a situation where race-based decision making is not rational, and will in general lead to suboptimal outcomes (in this case, having a juror biased against the prosecution).

~~12~~ 6 5 ~~10~~ n-Angry Men

I have reservations as to the wisdom—as well as the necessity—of Mr. Justice Blackmun's heavy reliance on numerology derived from statistical studies.

—ASSOCIATE JUSTICE LEWIS POWELL, CONCURRING OPINION
IN *BALLEW V. GEORGIA* (1978)

Two investigators from the Fulton County Solicitor General's Office went to the Paris Adult Theater in Atlanta, Georgia, on November 9, 1973, where they viewed the film *Behind the Green Door*. A few days later, they returned with a warrant, viewed the film again, then seized it, arresting manager Claude Ballew and a cashier. On November 26, investigators returned to the theater and viewed the film a third time; they obtained another warrant and returned on November 27, viewed the film a fourth time, and seized a second copy. Ballew was charged with two counts of distributing obscene material—a misdemeanor.

By Georgia state law, the Criminal Court of Atlanta tried its cases in front of a jury of five persons. Although Ballew requested that the case be transferred to the Superior Court (which would use a jury of twelve), the request was denied and Ballew was convicted. He appealed on the grounds that the small jury violated his 6th Amendment rights, and *Ballew v. Georgia* (1978) wound its way to the U.S. Supreme Court.

Downsizing the Jury

Although the 6th Amendment specifies a jury for "all criminal prosecutions," it is generally interpreted to apply only to felonies. Thus parking and speeding tickets can be decided by a bench trial, where guilt or innocence is decided by a judge; in fact, bench trials are probably the most common types of trials.

A traditional American jury for a criminal trial consists of 12 persons, all of whom must agree to the verdict: failure to reach unanimity results in a hung jury and a mistrial, requiring the whole process to begin anew. (Legally, a mistrial means that the trial never occurred.) However, the 6th Amendment only guarantees the right to a jury trial in federal cases; it specifies neither the size of the jury nor the quota by which it must reach a decision, nor does it restrict what the states can do.

For example, in 1967 Florida passed a law permitting the use of six-person juries for noncapital offenses. Such a jury convicted Johnny Williams of robbery, but he appealed on the grounds that the small jury violated his 6th Amendment rights. The Supreme Court considered the case in *Williams v. Florida* (1970). To decide the case, the Supreme Court considered the purpose of the 6th Amendment. In *Duncan v. Louisiana* (1968), the Court opined that the purpose of a jury trial is to prevent oppression by the government. It elaborated in the *Williams* decision:

> The performance of this role is not a function of the particular number of the body that makes up the jury. To be sure, the number should probably be large enough to promote group deliberation, free from outside attempts at intimidation, and to provide a fair possibility for obtaining a representative cross-section of the community. But we find little reason to think that these goals are in any meaningful sense less likely to be achieved when the jury numbers six, than when it numbers 12—particularly if the requirement of unanimity is retained. (399 U.S. 78, p. 100)

Thus the Court upheld the constitutionality of a six-person jury that rendered verdicts by unanimous votes.

The decision in *Williams* inspired research to quantify the effects of reducing jury size. These efforts fell into two categories. First, there were empirical studies, based on cases decided by six-person juries. Second, there were theoretical studies, based on models of how jurors voted.

The empirical studies are problematic for a number of reasons. First, prior to Florida's switch to six-person juries, the only cases routinely tried by smaller juries were civil cases. Such cases obviously have a different character than criminal trials. More importantly, civil trials have a much more lenient standard of proof: in contrast to the "proof beyond a reasonable doubt" of criminal trials, civil trials require only a "preponderance of evidence." Using data on six-person civil juries to assess six-person criminal juries is questionable.

Better results could be obtained by presenting the same evidence to juries of different sizes. But if the evidence is too clear-cut, both juries will return the same verdict, so useful information can be obtained only when the juries disagree. In such a case, we know that one jury has decided correctly and the other has not—but which one?

For this reason, empirical tests often use mock trials, where the guilt or innocence of the defendant is known. However, this approach also has its difficulties. Again, if the evidence is too clear-cut, both juries will return the same verdict; indeed, the Court considered and dismissed experiments of this sort in *Ballew*. But if the evidence is too ambiguous, a prosecutor might not risk trial, and instead offer a plea bargain or even dismiss the charges. Creating a suitable test case is very difficult.

Here we consider a more theoretical approach. Two analyses of significance appeared after *Williams*. The first, by Herbert Friedman of the College of William and Mary, approached the problem as follows. Let any given defendant have an *appearance of guilt*, which we can take as a measure of how guilty the defendant appears to the jurors. We can then plot the appearance of guilt against the probability of conviction to produce a *conviction rate curve*.

The value of this approach is that it provides us with a normative model: in other words, what we *hope* to see. There are two extremes. First, there is a *police state* (Friedman compared it to the Gestapo or GPU), where even the slightest appearance of guilt sufficed for a conviction. In contrast, Anglo-Saxon jurisprudence renders convictions unlikely until there is a substantial appearance of guilt. These extremes are shown in figure 4.

How can we evaluate juries? Suppose the appearance of guilt corresponds to the probability p that an individual juror finds a defendant guilty. We assume, for now, that the jurors decide independently. Then by the multiplication principle, the probability that all members of a 12-person jury find the defendant guilty is p^{12}, while the corresponding probability for a 6-person jury is p^6. This allows us to plot the predicted conviction rate cures for

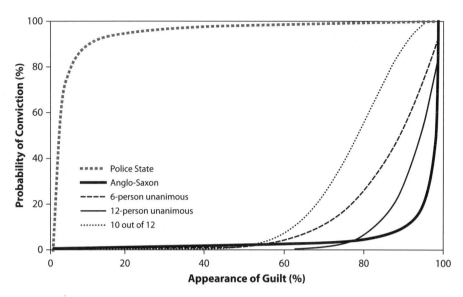

Figure 4. Conviction Rate Curve

6-person and 12-person juries. Friedman also considered the probability that a jury might convict a defendant with a non-unanimous verdict of 10 or more out of 12.

Clearly all these juries hew closer to the Anglo-Saxon ideal than to the police state, though the non-unanimous jury is farthest from the ideal and closest to the police state. But how can we use our curves? One possibility is that any jury system *closer* to the Anglo-Saxon ideal than those allowed by the Supreme Court should be permissible, while any jury system *farther* from the ideal than those rejected by the courts should be prohibited.

One month after Friedman's article appeared, the Court removed the unanimity requirement in *Apodaca v. Oregon* (1972) and *Johnson v. Louisiana* (1972). In both states, juries consisted of 12 persons, but could convict by a non-unanimous vote: 10 out of 12 in Oregon, and 9 out of 12 in Louisiana. The Court did not reference Friedman's work, but instead based the acceptability of non-unanimous verdicts on the fact that James Madison's original version of the 6th Amendment included the requirement that juries decide unanimously. This language was removed from the final version. As we noted earlier, one interpretation of the Constitution holds that omissions are just as significant as inclusions; in this case, since language requiring unanimity was omitted, we might hold that unanimity is not constitutionally required.

The second important theoretical study of jury size was conducted by two political scientists, Stuart Nagel and Marion Neef, in 1975. Nagel and Neef improved on Friedman's analysis by considering two types of defendants: those who are in fact guilty, and those who are in fact innocent. It is reasonable to assume that the probability of a juror voting to convict a guilty defendant is not the same as the probability of a juror voting to convict an innocent one; let these probabilities be p and q, respectively.

How can we find p and q? Nagel and Neef suggested that guilty defendants might be convicted at a 70% rate, so $p^{12} = 0.70$ and thus $p \approx 0.971$. Likewise, they assumed that innocent persons might be convicted at a 40% rate, so $q^{12} = 0.40$ and thus $q \approx 0.926$.

Next, they assumed that 95% of all defendants were in fact guilty. Suppose we consider 1,000 defendants. It follows that this group will consist of 950 actually guilty defendants, and 50 actually innocent defendants. Of the 950 guilty defendants, 70% of them (665) will be convicted and 30% (285) will be acquitted. In contrast, 50 of the defendants will in fact be innocent, and 40% of them, or 20, will be convicted.

Whether or not we find this objectionable is a question of legal philosophy. However, there is a dictum in Anglo-American law that "It is better to let ten guilty men go free than to convict one innocent." This originates from William Blackstone's *Commentaries on the Laws of England* (1752), and has been repeatedly cited by the Supreme Court as an ideal to which the justice system should aspire. Under the above assumptions, our 12-person jury sets free about 14 guilty defendants for every innocent convicted.

The problem is that it's not clear how to apply Blackstone's maxim. If we lower the standard of proof, we would acquit fewer guilty, but convict more innocents. While this would move us closer to the 10 to 1 ratio, do we want to convict more innocents? Alternatively, we might raise the standard of proof. This would acquit more innocents, which is desirable, though we would also acquit more guilty defendants. Again, while this might move us closer to the 10 to 1 ratio, it's not clear that we would want to do this.

Nagel and Neef used the following approach. Blackstone's maxim implies that the harm inflicted by convicting one innocent defendant is more than the harm inflicted by acquitting ten guilty defendants. If we set the harm of acquitting one guilty defendant at one unit, then the harm of convicting a guilty defendant is at least 10 units. Consequently, the harm done by our 12-person jury system, over 1,000 defendants, will be at least $285 \times 1 + 20 \times 10 = 485$ units.

In contrast, consider a 6-person jury. Under the same assumptions, the probability it will convict a guilty defendant will be $(0.971)^6 \approx 0.837$, and the probability it will convict an innocent defendant will be $(0.927)^6 \approx 0.632$. Thus in 1,000 defendants, 95% of whom are in fact guilty, there will be about 155 guilty set free, against 32 innocents convicted. The harm in this case would be $155 \times 1 + 32 \times 10 = 475$ units. Because the harm is less, this implies that a 6-person jury is actually better than a 12-person jury: while it convicts more innocents, it acquits fewer guilty, and at a tradeoff rate of 10 to 1, the increase in innocents convicted is more than paid for by the decrease in guilty acquitted. In fact, juries in the range of six to eight persons produced the least harm under these assumptions.

Justice Blackmun, writing the opinion in *Ballew*, found Nagel and Neef's conclusion convincing. The rest of the Court agreed that five-person juries violated the 6th Amendment, though Justice Powell, joined by Burger and Rehnquist, objected to the use of empirical and theoretical studies to make the decision, leading to the quote that opened this chapter. Powell's remark has been taken as indicative of judicial hostility to statistics. But in fairness, the conclusion reached by Nagel and Neef *only* holds because of a number of assumptions, all of which are purely conjectural, and the Court has consistently objected to purely hypothetical arguments.

To begin with, Nagel and Neef assumed that 95% of defendants were in fact guilty. However, the fraction of guilty defendants is unknown and, in principle, unknowable (since if we knew whether or not a defendant was in fact guilty, we could dispense with the jury trial). Moreover, even if the rate were "around" 95%, the slightest variation in the fraction of guilty defendants will lead to dramatically different results. For example, if the percentage of guilty defendants was 96%, the optimal jury size drops to one! On the other hand, if we decrease the percentage of guilty defendants to 90%, the optimal jury size skyrockets to twenty-two. Our knowledge of the rate at which innocent persons are put on trial is too uncertain to use a hypothetical value as the basis for any binding conclusions.

Next, Nagel and Neef assumed a conviction rate of $p = 70\%$ for guilty defendants and $q = 40\%$ for innocent defendants. However, we only know the total conviction rate. If it is in fact true that 95% of defendants are guilty, Nagel and Neef's assumptions lead to an overall conviction rate of 67%. This rate, at least, is consistent with actual conviction rates in many jurisdictions.

The problem is that there are an infinite number of combinations of p and q that lead to the same conviction rate, and different values of p and q yield very different conclusions regarding optimal jury size. If both innocent and guilty defendants are convicted at a 67% rate, this would give the observed 67% conviction rate, but our optimal jury size would drop to one! On the other hand, if guilty defendants were convicted at a 70% rate, innocent defendants would have to be convicted at a 10% rate in order to maintain a 67% conviction rate; however, under these assumptions, the optimal jury size is between 7 and 8.

Probability Distributions and the Law

Other objections can be raised against the Nagel and Neef approach, but the most common is a persistent belief that mathematics cannot be applied to predict human behavior. This is summarized in pithy comments like "Jurors don't flip coins." However, mathematical methods can successfully predict election outcomes, how crowds leave a stadium, and whether a person who likes *Don Quixote* would also like *Don Juan*, so there is no reason to deny that mathematics can be used to model human behavior. There are times when mathematics appears to produce badly flawed predictions. But when these cases are examined more carefully, the problems invariably stem from one of two sources. One problem is the use of a badly flawed mathematical model that no mathematician would accept as valid. Second, a model is only as good as the data fed into it: "garbage in, garbage out." Moreover, the probabilities do not model *how* the decisions are made, but rather the decisions themselves.

A more troublesome feature of the Nagel and Neef approach is how dependent their conclusions are on the assumed probabilities. As we saw, slight variations in the percentage of guilty defendants, or the conviction rates of innocent and guilty defendants, can lead to very different conclusions regarding the size of the jury. Fortunately, generalization is the heart and soul of mathematics, and if our conclusion depends on the probability we choose, we can use mathematics to choose *all* of them.

We can do this by using a probability distribution. Consider Friedman's approach, where a given defendant has some appearance of guilt. The probability that a particular defendant is convicted depends on their appearance of guilt. But we cannot predict, until the person is actually brought before a jury, what their appearance of guilt will be: thus the appearance of guilt is a random variable, and we can speak of its distribution among the possible defendants.

The tools of calculus allow us to handle any probability distribution we want, but we can illustrate the basic approach using a *discrete probability distribution*. Suppose suspects fall into one of four categories, A, B, C, and D, where the appearance of guilt is 10%, 40%, 70%, and 99% respectively. The prosecutor has several choices. Cases unlikely to yield convictions might be dismissed entirely; other cases might be sent to trial; but still others might be settled via a *plea bargain* which, if accepted, would produce a conviction on a lesser charge without a jury trial.

In this case, suppose our prosecutor dismisses all cases from category A (where the appearance of guilt is only 10%). Of the remainder, the prosecutor might choose to offer a plea bargain to persons from categories B and C (where the appearance of guilt is 40% and 70%, respectively): while there is evidence against these defendants, it is sufficiently weak that either side risks losing if the case goes to trial. Finally, since cases against category D defendants are very strong, the prosecutor might simply take the case to trial directly.

As a result of decisions like these, the defendants actually brought before a jury will be distributed in some fashion among the possible values of the appearance of guilt. For illustrative purposes, suppose that 5% of defendants have appearance of guilt 40%; 10% have appearance of guilt 70% (these corresponding to those who rejected a plea bargain offer); and 85% have appearance of guilt 99%. We can determine the fraction of convictions as follows.

To make sense of the numbers that follow, it's convenient to assume that we try a large number of cases; to extend Nagel and Neef's analysis, suppose we consider the results of 1,000 cases. Under our assumptions, 5% of 1,000 defendants, or 50, will have appearance of guilt 40%; 100 will have appearance of guilt 70%; and 850 will have appearance of guilt 99%.

What happens to these defendants? First, a defendant who has appearance of guilt 40% has probability $(0.40)^{12} \approx 0.00001678$ of being convicted. Thus this group (of 50) will return $50 \times 0.00001678 \approx 0.000839$ guilty verdicts: effectively, none of them will be convicted.

Next, consider the 100 defendants who have appearance of guilt 70%. The probability of conviction for these defendants is $(0.70)^{12} \approx 0.01384$ will be convicted, so they will contribute $100 \times 0.01384 \approx 1.384$ convictions.

Finally, the 850 defendants with appearance of guilt 99% will be convicted at a rate of $(0.99)^{12} \approx 0.8864$, for a total of $850 \times 0.8864 \approx 753.4271$ guilty verdicts. Altogether, there will be $753.4271 + 1.384 + 0.000839 \approx 755$ convictions, for an overall conviction rate of about 76%.

It's worth emphasizing that our results will depend on the legal culture of a given jurisdiction, and to some degree the individual prosecutors and defense attorneys. For example, if the court docket is very full, the prosecutor might offer (and the defendant would take) plea bargains for those with a 99% or higher appearance of guilt; thus we'd see a much smaller fraction of such defendants being brought before a jury. Or the prosecutor might simply refuse to offer plea bargains at all, so we'd see an even higher fraction of defendants with a 40% or 70% appearance of guilt.

It might seem that this makes analyzing jury systems on a theoretical basis impossible, since what holds in one jurisdiction will fail to hold in another with a different probability distribution. But this is one case where the legal system makes it easier to perform a quantitative analysis. Because the law operates on precedent, if we prohibit a jury type in one jurisdiction, we threaten the existence of that jury type in all jurisdictions. There are exceptions to this general rule if the jurisdictions are sufficiently different. For example, the juvenile court system is not required to provide trial by jury.

This frees us to approach the problem as follows. Suppose we analyze a type of jury under a given probability distribution. We will find the results either palatable or objectionable. If we find them palatable, the jury problem deserves more study as to whether the given probability distribution is representative of actual jurisdictions. But if we find the results objectionable, then if the probability distribution corresponds to *any* jurisdiction, the jury type under consideration should be forbidden pending further investigation. (As we pointed out in the chapter on jury discrimination, *systems* do not have a presumption of innocence.) Mathematics might not tell us what we should do—but it can at least tell us what to avoid.

SAM and Jury Verdicts

This raises a new problem: What is objectionable? Obviously, if a jury system too often convicts the innocent or acquits the guilty, it should be rejected. However, since the jury is the instrument by which society determines the guilt or innocence of a defendant, we will never know if it has in fact convicted the innocent or acquitted the guilty. This conundrum leads to the Sample Accuracy Model (SAM), proposed by George C. Thomas III and Barry S. Pollack in 1992.

SAM focuses on whether the jury returns verdicts that are acceptable to the general society. This might seem a peculiar viewpoint to take, especially if one takes the view that a defendant is in fact either guilty or innocent regardless of what society believes. However, it is a pragmatic viewpoint: if the legal system too often produces results at odds with popular sentiment, individuals will turn to alternative methods of dispensing justice.

Thomas and Pollack approached the problem as follows. Consider the jury as a sample of society, to be used to determine the sentiment of society. Then we are interested when the verdict rendered by the jury differs from that of society. We might simply consider the cases where the jury renders one decision and society renders another. However, we can consider a more nuanced approach as follows.

We'll say that a verdict is *contested* if society reaches a different conclusion after the jury's verdict, and *wrongful* if society reaches its verdict before the jury. There are four probabilities of importance:

- The probability of a contested conviction, where society acquits someone the jury has convicted
- The probability of a contested acquittal, where society convicts someone the jury has acquitted
- The probability of a wrongful acquittal, where the jury acquits someone that society has convicted
- The probability of a wrongful conviction, where the jury convicts someone that society has acquitted

Which of these probabilities is most important? For that, we should consider some recent contentious cases.

In 1991, Troy Davis was convicted of the murder of Georgia police officer Mark MacPhail and sentenced to death. A grassroots movement arose to overturn the conviction and death sentence. The movement ultimately failed, and Davis would be executed in 2011. In terms of the above probabilities, the Davis case is a contested conviction: at the instant of his conviction, no one knew the sentiment of the general public.

In contrast, consider the 1992 acquittal of Los Angeles police officers charged with beating Rodney King in 1991. Continuous media coverage made the sentiment of society clear, and in this case the verdict sparked riots that devastated large areas of the city of Los Angeles. Here it would be reasonable to focus on the probability of a wrongful acquittal. More recently, the 2013 acquittal of George Zimmerman, charged with the murder of Trayvon Martin in 2012, and the 2011 acquittal of Casey Anthony, charged with the murder of her two-year-old daughter, also offer cases where the probability of interest is that the jury acquits, given that society has convicted.

Based on these examples, it's unclear which of the four probabilities should be considered. Fortunately, it's easy enough to compute *all* of them.

How do we find these probabilities? To begin with, consider a simpler situation: Suppose we have a group of 50 men and 50 women; we know that of the group, 35 have red hair, and 15 members of the group are redheaded men.

From this information, we know the prior probability that a person has red hair is $35/100 = 35\%$. But as we saw earlier, Bayes's theorem allows us to revise this probability if we have more information. Thus, suppose we knew that the person in question was male: what is the probability that a person has red hair, given that we know the person is male?

We might reason as follows. Since we know the person is male, we can focus on the 50 men in the group. Since we also know there are 15 redheaded men in the group, it follows that 15 of those 50 will be redheaded. Thus the probability that a person has red hair, given that we know the person is male, is $15/50 = 30\%$. If we let $R = $ a person has red hair, and $M = $ a person is male, then this reasoning can be summarized as: $P(R \mid M) = P(R \text{ and } M) \div P(M)$.

It follows that if we want to find the probability of a contested conviction, which is to say the probability that society acquits given that the jury has convicted, then we want to find the probability that society acquits *and* the jury convicts, and divide it by the probability that the jury convicts.

There is an easy transition from Friedman's work to SAM if we take the appearance of guilt as the fraction of the general population who view a defendant guilty. For our illustrative discrete probability distribution above, we found that the probability that a jury convicts a defendant is 0.7548.

What about the probability that society acquits and the jury convicts? In our example, defendants fell into one of three categories: those with a 40% appearance of guilt; those with a 70% appearance of guilt; and those with a 99% appearance of guilt. Following the majoritarian viewpoint, only the first group would be acquitted by society; we determined the probability that a member of this group would be convicted by a jury is 0.00001678. It follows that the probability that society acquits a defendant that the jury has convicted will be

$$\frac{0.00001678}{0.7548} \approx 0.00002223$$

or about 1 in 45,000 convictions.

What if we switch to a 6-person jury? Again for our illustrative probability distribution, we find the probability that such a jury convicts a defendant will be 0.8122, and the probability that a defendant is acquitted by society but convicted by the jury is 0.004096. Then the probability that society would acquit a defendant convicted by the jury is

$$\frac{0.004096}{0.8122} \approx 0.005043$$

or about 1 in 200 cases. Under these assumptions, a 6-person jury is about 226 times more likely to return a contested conviction than a 12-person jury.

What about the 12-person juries that don't require unanimity? The probability of a contested conviction by an *Apodaca*-type jury, where 10 out of 12 jurors are needed to agree to a verdict, is 0.003014, while a *Johnson*-type jury, where 9 out of 12 jurors are needed to agree to a verdict, has a probability of a contested conviction of 0.01696. Finally, the probability of a contested conviction by a *Ballew*-type jury, where 5 out of 5 jurors must agree to a verdict, is 0.01240.

If we take the viewpoint that society determines what the true verdict should be, then these probabilities correspond to the false alarm rate in statistical decision theory. While they all fall below the 5% rate commonly accepted, remember that this rate depends on the consequences of a false alarm. In this case, a false alarm is (from society's point of view) a wrongful conviction, and we should regard them as especially serious. Thus a lower false alarm rate is warranted. A commonly used lower rate is 1%, and we note that both the *Ballew*-type jury and the *Johnson*-type jury exceed this rate; hence we might view this type of jury with some trepidation. During the 1970s, Louisiana changed its laws and, like Oregon, now requires 10 out of 12 jurors to agree to a verdict; we may view this as a move in the direction of greater justice.

About a year after *Ballew*, the Supreme Court heard the case of *Burch v. Louisiana* (1979). As in *Ballew*, the case concerned the showing of a motion picture deemed obscene. This time the defendant was tried by a six-person jury where conviction could be secured by a vote of five or more jurors. On April 17, 1979, the Court rendered its decision: "Lines must be drawn somewhere if the substance of the jury trial right is to be preserved." Consequently the Supreme Court ruled that while juries of six persons were constitutional, the unanimity requirement had to be retained for such juries.

The Supreme Court's judgment was based largely on intuition and legal tradition. It is therefore rather surprising that it turns out to be mathematically consistent. As above, we can calculate the probability of a contested verdict by a 6-person jury where 5 out of 6 is sufficient for conviction: the probability is 0.04588, which not only exceeds the 1% false alarm rate we aspire to for juries, but approaches the 5% false alarm rate that is the highest generally acceptable.

Of course, these computations are based on a specific discrete probability distribution, and so might be regarded as too hypothetical to be of any value. But, as we noted, if *any* jurisdiction had such a probability distribution, the calculations above would suggest forbidding juries of particular types for that jurisdiction; and precedent would establish a strong case for forbidding those juries everywhere.

To make a stronger case, we might make use of a probability distribution deemed more likely to be representative. Thomas and Pollack themselves analyzed the problem using a *uniform probability distribution*, assuming that the appearance of guilt of defendants was evenly distributed between 20% and 100%. Analysis of this case is not different in principle from the case discussed previously, though in practice it requires calculus. The probabilities of wrongful and contested convictions and acquittals are shown in table 13. If we look at the table carefully, we note one remarkable feature: the Supreme Court's decisions are mathematically consistent. In particular, the permitted juries have lower probabilities of contested convictions, contested acquittals, wrongful convictions, and wrongful acquittals than the forbidden juries.

Triple Jeopardy

It appears that the Court's decisions in *Williams* and *Apodaca* are vindicated mathematically: 6-person juries, as long as they retain the unanimity requirement, do not perform substantially worse than 12-person juries. However, there are other factors to consider before wholehearted approval of 6-person juries or permission of non-unanimous decision rules.

Table 13. Probabilities of Jury-Society Disagreements

Jury Size and Quota	Probability Contested Conviction	Probability Contested Acquittal	Probability Wrongful Conviction	Probability Wrongful Acquittal	Legal Status
12 out of 12	0.000122	0.00222	0.0000313	0.0000188	Permitted (traditional)
6 out of 6	0.0078	0.03725	0.00371	0.0022321	Permitted (*Williams v. Florida*, 1970)
10 out of 12	0.00435	0.01653	0.0033	0.00201	Permitted (*Apodaca v. Oregon*, 1972)
5 out of 5	0.0156	0.05960	0.0086	0.00521	Forbidden (*Ballew v. Georgia*, 1978)
5 out of 6	0.0350	0.08941	0.0333	0.0201	Forbidden (*Burch v. Louisiana*, 1979)

One problem with smaller juries is that they are less likely to be diverse. The jury that acquitted George Zimmerman consisted of six persons—all women. Jury consultants allege to be able to select juries that will be more likely to render a given verdict, but it's not clear that this is the case: while a jury of six women did in fact acquit Zimmerman, it's not clear that a differently constituted jury would have rendered a different verdict. As with empirical studies of jury performance, the only way to gauge the effectiveness of a jury consultant's decisions is to put identical cases before different juries.

A bigger problem is that smaller juries and non-unanimous decision rules may exacerbate the threat posed by biased jurors. Consider a jury where one or more members are biased against the defendant. For example, when confronted with a defendant with an appearance of guilt of 40%, SAM assumes that a juror also has a 40% likelihood of voting to convict. However, we could easily imagine a biased juror who is more than 40% likely to vote to convict.

Suppose we assume that these biased jurors are also distributed uniformly over the remaining interval. In other words, if the defendant has an appearance of guilt of 40%, then the juror drawn from the general population will have a 40% chance of voting to convict, but the biased juror will have some probability between 40% and 100% of voting to convict. Under these assumptions, we find that if a 12-person jury requiring unanimity has one biased juror, the probability that the jury convicts someone society has acquitted rises to 0.134%. By a similar computation, we find that the 6-person jury with unanimity and one biased juror returns a wrongful conviction about 5.1% of the time, and a 12-person jury with one biased juror that can return a verdict by a vote of 10 out of 12 returns a wrongful conviction about 2.3% of the time.

Clearly, the 6-person unanimous and 12-person non-unanimous juries are more strongly affected by the presence of biased jurors. The magnitude of this effect is surprising: even with 6 biased jurors, the probability that a 12-person jury that requires unanimity will return wrongful convictions is only about 1.5%.

Let us say that the jury is biased if it includes enough biased jurors to raise the probability of a wrongful conviction to more than 1%. (We focus on wrongful convictions because the probability of a contested conviction is much lower.) The 6-person jury only needs one biased juror to become a biased jury; the 12-person jury where a vote of 10 suffices to convict will also be biased when it includes one biased juror. In contrast, the 12-person jury can have up to five biased jurors and still have a low probability of returning a wrongful conviction.

The effect of bias is further magnified if we consider the rationale behind smaller juries: smaller juries require smaller venires, and thus impose a smaller societal cost. Suppose that half of those summoned for jury service are excused or dismissed for cause. Thus to obtain a jury of 12, the initial venire would need to contain 24 persons. In contrast, a jury of 6 would only require a 12-person venire.

However, the existence of peremptory challenges increases this number. For example, suppose prosecution and defense both have six peremptory challenges. For a 12-person jury, this means that we must have 24 persons not excused or dismissed for cause, and thus (under our earlier assumptions) we must summon 48 persons for jury service.

What about 6-person juries? If both sides still have six peremptory challenges, this means that we must have 18 persons not excused or dismissed for cause, and we must summon 36 persons initially. Thus, while the jury size is halved, failure to reduce the peremptory challenges by the same fraction means that we only reduce the number of persons summoned by 25%.

But remember, the purpose of peremptory challenges is to remove jurors who might be biased against one side or the other. By reducing the number of peremptory challenges, we limit the ability of the defense to strike biased jurors. How much does the reduction in the number of peremptory challenges affect the ability to produce an impartial jury?

One approach to this is the following, applied to a system where the jury size is 12 and each side has 6 peremptory challenges. This system requires that we have $12 + 6 + 6 = 24$ jurors who survive voir dire. Let p be the probability that one of these jurors who survives voir dire is biased. We can compute the probability that this group of 24 contains any number of biased jurors, from 0 to 24.

Since the defense has 6 peremptory challenges, it follows that if there are 6 or fewer biased jurors, none will make it onto the jury itself. So, what is the probability that the 24 remaining jurors will include 7 or more biased jurors? If we wish to compare 12-person juries to 6-person juries, we need to go further. Since a 12-person jury can have up to 5 biased jurors before its contested conviction rate rises to worrisome levels, it follows that the venire, after voir dire, must contain at least 12 biased jurors: 6 of these will be struck peremptorily, allowing the remaining 6 onto the jury itself.

We can compute the probability that 12 or more persons in the 24-person venire are biased. Naturally, this will depend on the fraction of the population that is biased: for example, if 5% of the population is biased, then the probability that a 24-person venire has at least 13 biased persons is about 1 in 3 billion. We can use this as the probability that the 12-person jury is biased, when unanimity is required for conviction and each side has 6 peremptory challenges.

What about the 6-person jury? If each side has 3 peremptory challenges, then the venire must consist of $6 + 3 + 3 = 12$ persons. Since even one biased juror on a 6-person jury can profoundly affect the probability of a contested conviction, it follows that if there are 4 biased jurors in the venire, at least one will make it onto the jury itself. Thus the probability that a 6-person jury is biased is the same as the probability that the venire of 12 includes 4 biased jurors. Again, assuming 5% of the population biased, we can determine this probability: It is about 1 in 500.

What if we don't reduce the number of peremptory challenges? Because of the greater sensitivity of the smaller jury to the presence of biased jurors, we need at least 13 peremptory challenges per side to reduce the likelihood of producing a biased jury to that equivalent to a 12-person jury with 6 peremptory challenges per side. Thus in contrast to the 12-person jury that requires a venire of 24 before peremptory challenges, the 6-person jury requires a venire of *32* before peremptory challenges. Rather than reducing the required size of the venire, smaller juries increase them.

We can do a similar computation with the 12-person jury that only requires 10 out of 12 to convict. As with the 12-person jury requiring unanimity, let us suppose that each side has 6 peremptory challenges, so a venire of 24 must remain after voir dire. Since only one biased juror is necessary to produce a biased jury in this case, the jury will be biased if, after voir dire, the venire contains 7 or more biased jurors. Again, assuming that 5% of the population is biased, we find this probability to be about 1 in 8000.

Figure 5 shows the probability of obtaining a biased jury for different jury sizes and different probabilities of summoning a biased juror. We assume that for the 12-person juries, each side has six peremptory challenges, and for 6-person juries, each side has three. Clearly a defendant facing a 12-person jury requiring unanimity is better insulated from the presence of biased jurors than defendants confronting other types of juries. The difference becomes particularly dramatic when 25% of the post–voir-dire venire is biased: the probability that a 6-person jury is biased rises to about 1 in 6; for the 12-person non-unanimous jury, the probability is about 1 in 10; and for the 12-person unanimous jury, the probability is 1 in 1,000.

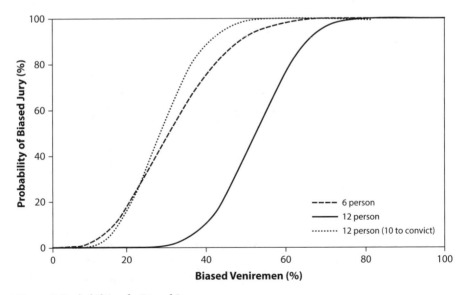

Figure 5. Probability of a Biased Jury

It would seem that smaller juries and non-unanimous verdicts pose a triple threat to the perception of justice. First, they are more likely to produce contested verdicts. Second, they are more sensitive to the inclusion of one or more biased jurors. Third, there is an increased probability that biased jurors will make it on to the jury panel.

Deliberating Deliberation

There are two rather significant omissions in the preceding analysis. The first is relatively minor: the assumption that all jurors had the same likelihood of voting to convict or to make the right decision. However, as our computations using biased jurors show, we can easily modify the analysis to allow for differences in the likelihood that a juror makes the correct decision.

A more problematic omission is using p^n as the probability that an n-person jury votes unanimously to convict. This assumes that jurors vote independently, an assumption we may justify to some extent on the grounds of tractability. But it also assumes that jurors only vote once—which is at odds with how real juries operate. If the initial vote does not meet quota, jurors must deliberate and another vote must be taken somewhat later.

Condorcet considered the jury size and quota problem in *Essay on the Application of Analysis to the Probability of Plurality Voting* (1785). His analysis began with the notion that any given person has some probability p of making the correct decision, and complementary probability $1-p$ of making the wrong decision. (Condorcet actually used v and e for these probabilities, from the French words *vérité* ("truth") and *erreur* ("error").) To begin with, suppose that there are n jurors and that the decision must be unanimous. If we suppose that each juror's decision is independent of the others, then the probability that a unanimous correct decision is reached on the first ballot is p^n, while the probability of a unanimous incorrect decision is $(1-p)^n$. Hence the probability that the first ballot will return a decision will be $p^n + (1-p)^n$, so the probability that the decision is correct, given that it is rendered on the first ballot, is $\dfrac{p^n}{p^n + (1-p)^n}$.

If the jury does not reach a decision on the first ballot, it must conduct a second one. Condorcet assumed that the probability of making a correct or incorrect decision doesn't change, an assumption Condorcet acknowledged as unrealistic; he promised to return to it later (but never did). Under these assumptions, we come to a surprise result: the probability that the jury *eventually* reaches the correct decision is *also* $\dfrac{p^n}{p^n+(1-p)^n}$! As Condorcet pointed out, it means that the probability of *eventually* returning the correct verdict is the same as the probability they will render the correct verdict on the first ballot.

One of the important consequences of this work is that if n is large enough, the jury is likely to render a correct decision even if the individual jurors are almost as likely to make the wrong decision as the right decision. For example, suppose Congress had to render its decisions on legislation unanimously, so we have $n = 435$. If $p = 0.505$, the probability that the correct decision would eventually be reached is 99.98%. This lends weight to the notion of collective wisdom, and has been called the fundamental theorem of democracy (though today we might call it the fundamental theorem of crowd sourcing).

Condorcet himself was more cautious, since the reverse occurs if $p < 0.5$. In this case, the crowd is almost certain to reach the incorrect decision. Moreover, "The assumption that [$p < 0.5$] is not absurd, as there are a great many complex and important questions subject to the realm of prejudices and passions, upon which it is likely that an uneducated man will hold an erroneous opinion."[1]

What happens when we incorporate the Condorcet model into SAM? If we use $\dfrac{x^n}{x^n+(1-x)^n}$ as the probability a n-person jury convicts a person whose appearance of guilt is x, we find that 12-person juries return contested convictions about 3.4% of the time, while 6-person juries return contested convictions about 6.1% of the time, and a 12-person jury that returns a verdict by a vote of 10 or more has an 8.5% probability of returning a contested conviction. Again, the 12-person unanimous jury provides better protection against contested convictions than any other type of jury in common usage.

A Jury of One Peer

The biggest objection to Condorcet's model is that it does not take into account juror interactions. In effect, it treats jury decision-making as a gigantic

1. Condorcet, *Essay on the Application of Analysis to the Probability of Plurality Voting*, pp. 7–8.

game of rock-paper-scissors, played over and over again until all players make the same decision. But one might expect that in a real jury situation, a juror who is outvoted would tend to reevaluate his or her decision on the next ballot.

The problem is that deliberation literally occurs behind closed doors, so gaining insight into how jurors decide cases is extremely difficult. One important study emerged in the 1960s under the auspices of the Chicago Jury Project. After some wrangling, project researchers won permission to record a half-dozen federal jury proceedings. However, after one of the judges involved played an edited recording of jury deliberation at a conference, angry denouncements over "bugging" jury deliberation ensued, leading to a federal law banning such recordings in the future. Subsequent jury research has had to use indirect methods, such as post-verdict interviews of jurors. In spite of these obstacles, Harry Kalven, Jr. and Hans Zeisel completed and published *The American Jury* in 1966. It remains one of the more comprehensive studies of the jury decision process.

We can use the Kalven and Zeisel study as a starting point for a theoretical analysis of jury decision-making. For example, jury decision-making models require the probability that a juror makes the correct decision. The obvious way to do this is to compare a juror's decision with the correct one—but this requires that we know what the correct decision is, and if we knew that, a jury would be superfluous.

As it turns out, we can find the probability *without* knowing the correct decision. This is because any decision requires a certain number of jurors to vote one way or the other. For example, the 12-person criminal juries studied by Kalven and Zeisel had to return unanimous verdicts. Suppose a verdict is returned on the first ballot. Then either all 12 jurors made the correct decision, with probability p^{12}, or all 12 made the incorrect decision, with probability $(1-p)^{12}$. Kalven and Zeisel found about 31% of criminal cases were decided on the first ballot, so we can solve the equation $p^{12} + (1-p)^{12} = 0.31$ for the probability of a juror making the correct decision.

Inconveniently, there are two solutions: $p = 0.9070$ and $p = 0.0930$. Thus, either jurors are about 91% likely to make the correct decision—or 91% likely to make the wrong decision. While we agree with Condorcet that there are a great many questions about which an uneducated person will hold an erroneous opinion, it stretches the limits of pessimism to believe that they will hold an erroneous opinion 91% of the time.

At the same time, our ultimate conclusion stretches the limits of optimism: If we take $p = 0.91$, we find that a 12-person jury will be phenomenally accurate—less than one case in one *trillion* would be decided incorrectly. To put this number in perspective, the number of jury trials worldwide has been estimated at about 100,000 per year, most of which occur in the United States. Our computed probability translates into one incorrect verdict every ten million years, or longer than the human species has been in existence.

The obvious problem is that we have not taken into account juror interactions. We can do this using *transition probabilities*: the probability of going from one situation (say, a vote of 8 to 4 to convict) to another (12 to 0 to convict). Kalven and Zeisel's study suggests that if the initial vote favored one verdict, that verdict tended to prevail. Thus, in a study of 225 cases tried in Chicago and Brooklyn, 41 (18%) began with between 7 and 11 jurors favoring acquittal; of these cases, 37 (91%) eventually acquitted their defendants. Likewise, 105 (47%) began with between 7 and 11 jurors favoring conviction; of these, 90 (86%) eventually convicted their defendants. Finally, 10 cases (4%) began with a tie vote; of these, 5 convicted and 5 acquitted their defendants.

To illustrate how transition probabilities work, suppose we have a three-person tribunal. Let p be the probability that a member judge of the tribunal makes the correct decision (convict or acquit) regarding a case on the initial vote. There are three ways the initial vote can be cast. First, the vote can be unanimous and correct. The probability that this occurs is p^3 and since all the judges agree, this will be the final verdict. Thus, this situation contributes p^3 to the fraction of correct verdicts.

Next, the vote can be 2 to 1 favoring the correct verdict. The probability that this occurs is $3p^2(1-p)$. Kalven and Zeisel's work suggests that about 90% of the time, the final verdict is the one favored by the initial verdict; thus this situation contributes fraction $(0.90)\, 3p^2(1-p)$ to the correct verdicts.

It's also possible that the vote can be 1 to 2 favoring the correct verdict. In this case, the initial vote favors the *incorrect* verdict. However, it's still possible for the holdout to persuade the other two. For convenience, we'll assume that since the final vote matches the initial vote 90% of the time, there is a mismatch 10% of the time. While this precludes the possibility of a hung jury, we note that if the jury *is* hung, then legally the trial never occurred, which we may treat mathematically by prohibiting such an outcome. Under these assumptions, an initial vote favoring the incorrect verdict will lead to a final correct verdict in $(0.10)\, 3(1-p)^2 p$ cases.

Finally, the vote can be 0 to 3 favoring the correct verdict. The probability that all three jurors make the incorrect decision is $(1-p)^3$. But since they have rendered a unanimous verdict, the incorrect verdict stands, and when this occurs, they contribute no correct verdicts.

Putting these fractional parts together, we find that a 3-person tribunal returns the correct verdict in

$$p^3 + 0.9\,(3p^2)(1+p) + 0.1\,(3p)(1+p)^2$$

of the cases presented to it. We can construct a similar expression for the probability that a traditional 12-person jury arrives at a correct decision, again assuming that 90% of the time, the eventual verdict is that favored by the initial ballot. Since it's possible, with 12 jurors, for the initial ballot to be a tie, we must include another transition probability: again, Kalven and Zeisel's data suggest that in such cases, half the verdicts lead to the correct decision and the other half to the incorrect decision.

An easy way to compare the two jury systems is to graph the probabilities that they will produce a correct verdict. When we do so, we make a remarkable discovery, shown in figure there is only a small difference between the reliability of a 3-person tribunal and a traditional 12-person jury!

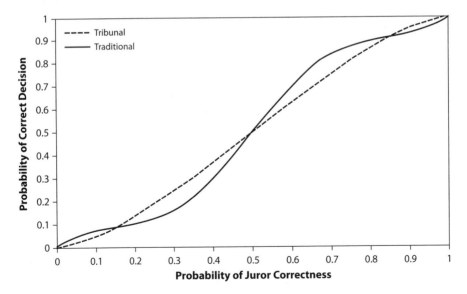

Figure 6. Jury Correctness (Transition Model)

We can go further. It seems reasonable to suppose that the closer the initial ballot is to a verdict, the more likely that verdict will be reached: if 11 out of 12 jurors vote to convict, it seems more likely that such a jury will eventually convict than one where only 7 of 12 jurors voted to convict. Kalven and Zeisel's study did not provide enough examples of such transitions to make any reliable assessment of these probabilities, but in 2008, Eric Helland and Yaron Raviv, two economics professors at Claremont McKenna College in California, considered the problem from a theoretical point of view. Under the assumption that the probability of a verdict is proportional to the initial ballot (so if the initial ballot was 11 out of 12 to convict, the probability of eventually obtaining a conviction is 11/12), they showed that in general, the probability that the *jury* makes the correct decision is equal to the probability that any single *juror* does so. In other words, a jury of 12 is no more reliable than a jury of one!

Of course, we may reasonably suspect that there are many topics "upon which it is likely that an uneducated man will hold an erroneous opinion" and doubt that a jury of one will satisfy the needs of justice; at the same time, it's worth noting that by far the most common type of trial is a bench trial, where guilt, innocence, and punishment are decided by one person. *Duncan v. Louisiana* reinforced a century of legal precedent by noting that bench trials sufficed for sufficiently minor crimes. However, the Supreme Court held that the severity of a crime should be determined by the severity of punishment. Thus speeding tickets and traffic violations, which incur only minor penalties, do not require jury trials.

Modeling Reality

The biggest challenge to creating a good model for the behavior of juries is accounting for group deliberation. SAM avoided the issue by focusing on the results of the first ballot only. It follows that as long as there *is* only one ballot, SAM can tell us a great deal about how juries perform. Thus we might require that, before jurors begin deliberation, they submit a ballot indicating their decision.

The problem is that if the first ballot does not render a verdict, deliberation must proceed, and our models are no longer accurate mirrors of reality. If jury size and quota were determined by the laws of physics, we would have to alter our models to fit reality. But jury size and quota are determined by the laws of man, which gives us the unique opportunity to alter reality to fit our models. In this case, we can change the jury size and quota so that a greater fraction of cases are decided by the first ballot.

Using the assumptions from SAM about the probability of a vote to convict, we find that a 12-person jury will render a first-ballot decision in roughly 10% of the cases, while a 6-person jury will render a first-ballot decision in 22% of the cases. The 12-person jury with a quota of 10 does even better: 36% of its cases will be decided on the first ballot. This supports one argument for smaller juries or non-unanimous quotas: they are able to render verdicts more quickly. We can reconcile these figures with the observation that about one-third of the cases tried by 12-person juries render a decision after one ballot by noting that the values stated assume no deliberation has occurred.

If we make rendering a verdict more quickly a goal, then it is worth considering *increasing* the size of the jury while *decreasing* its quota. For example, a jury of 18, where 16 are required to agree to a verdict, would render a first-ballot decision in about 18% of the cases—comparable to a 6-person jury. Moreover, such a jury would have the following probabilities of disputed verdicts:

- The probability of a wrongful conviction is 0.0001091.
- The probability of a wrongful acquittal is 0.000170.
- The probability of a contested conviction is 0.0001348.
- The probability of a contested acquittal is 0.0012100.

These compare favorably with the 12-person unanimous juries.

What about sensitivity to biased jurors? In order to produce a biased jury, the 18-person jury, with a quota of 16 required for a verdict, must have 7 or more biased jurors. Again, this compares favorably to the 12-person jury and is much better than the situation with the 6-person or non-unanimous jury.

Part of the rationale for decreasing the jury size is reducing the number of persons that must be summoned for jury service. Increasing the jury size obviously would require more persons to *serve* on juries, but if we reduce the number of peremptory challenges, we can keep the venire the same size. Instead of a 12-person jury panel, with each side having 12 peremptory challenges, which requires that we summon a minimum of 36 persons for each trial, we could have an 18-person jury panel with each side having 9 peremptory challenges, and still only have to summon 36 persons for each trial.

Because the 18-person jury with a quota of 16 is more sensitive to biased jurors than the 12-person unanimous jury, there will be a greater risk of obtaining a biased jury. Thus if 25% of the post–voir-dire venire is biased, the probability of obtaining a biased 18-person jury with a quota of 16 is about 1 in 85. While this is worse than the 1 in 1,000 probability for the traditional jury, it is definitely better than the 1 in 6 chance for the 6-person jury and the 1 in 10 chance for the 12-person non-unanimous jury.

Supporters justify smaller juries and non-unanimous decision rules by pointing out that they require summoning fewer persons for jury duty, and can render decisions more quickly. However, it's not clear that decreasing the size or quota of a jury will produce results consistent with our notion of justice. But if rendering decisions more quickly is desirable, and if the number of persons called for jury service is an issue, jurisdictions might well consider 18-person juries with a quota of 16 required to render a verdict: such juries would render decisions more quickly than 12-person unanimous juries, while maintaining roughly equivalent rates of disputed verdicts. Moreover, they need not require larger venires than 12-person unanimous juries, since a modest reduction in the number of peremptory challenges would still leave both sides ample opportunity to produce an unbiased jury.

The Peril and Promise of Social Network Analysis

Excessive bail shall not be required, nor excessive fines imposed, nor
cruel and unusual punishments inflicted.
—U.S. CONSTITUTION, AMENDMENT VIII

Theresa McIntyre Smith was convicted, on November 22, 2004, of conspiracy
to distribute more than 5 kilograms of cocaine. Smith's involvement came
through Roy Mercer, her hair stylist, whom she met in 1992. Mercer began
dealing drugs in 1995, and by 1997 had successfully built a cocaine distribution ring that involved several friends and family members. In the meantime,
Smith began working at Continental Airlines and, as one of her employment
perks, received "buddy pass" tickets that allowed friends and family members
to travel cheaply. She sold some of these to Mercer as favors; on occasion, she
also drove his friends and family to the airport.

On June 23, 1999, at the Detroit Metropolitan Airport, Smith identified a
bag belonging to one of Mercer's couriers. The bag contained 11 kilograms of
cocaine; this led to Smith's arrest and eventual conviction. Federal law mandates that all members of a conspiracy to manufacture and distribute drugs
receive a sentence of not less than 10 years. However, there is a "safety valve"
provision that allows reduced sentences for those who cooperate with the
prosecution or for first-time offenders. The couriers, who made upwards of
$2,000 a trip for as many trips as they cared to make, did cooperate, and
received between 12 and 24 months.

Smith claimed to have no knowledge of the conspiracy, a claim backed by
Mercer himself. Thus she could not "cooperate" with the prosecution by providing them with any information. As a result, she received the mandatory

minimum sentence of 10 years. In principle, the "safety valve" covers those who have no useful information to offer the prosecution. In Smith's case, the prosecution felt that Smith exhibited "willful blindness" to the conspiracy (in other words, she should have known an illegal enterprise was in operation), and was thus a knowing participant who refused to cooperate; this overrode the fact that she was also eligible for a reduced sentence by virtue of being a first-time offender.

Smith's case raises a number of questions, but one of particular significance is this: The very nature of a conspiracy means that most of the participants will be unaware of its details. Although Smith provided material support to Mercer's drug trade by providing his couriers with cheap tickets and transportation to and from the airport, it is well within the realm of possibility that she did so without knowledge of the conspiracy itself.

Even if Smith was aware of the conspiracy, as prosecutors allege, the principles of fairness and the 8th Amendment's prohibition of cruel and unusual punishment require that the punishment should be proportionate to the crime. So, then, was Smith, as prosecutors allege, a key figure in the conspiracy, worthy of a 10-year sentence? Or was she at most a peripheral figure, whose status as a first-time offender should have qualified for a reduction in sentence via the safety valve provision?

Social Network Analysis

In late 2013, outrage over the extreme sentences imposed on minor participants in drug distribution rings caused Attorney General Eric Holder to issue a directive that federal prosecutors not charge low-level nonviolent offenders with offenses that carried extreme penalties. But this directive does not affect those already given extended sentences and, since the laws mandating the extended sentences are still in force, Holder or a later attorney general could rescind the order and allow low-level nonviolent offenders to face 10-year sentences.

More broadly, a person convicted in the federal system is sentenced according to a set of guidelines that compute an offense level, based on the nature of the crime. Thus, those identified as organizers or managers could face a 3- or 4-level increase in the offense level, while those identified as minor participants could earn a 2- to 4-level reduction. Since the recommended sentence roughly doubles for every six offense levels, determining the extent of a person's involvement in a conspiracy is an important judicial task.

Mathematics can offer insight into this question through the use of *social network analysis*, which might be described as the geometry of organizations. Social network analysis begins with a *graph*, such as that in figure 7, which consists of a set of points A, B, C, etc., and lines joining some (but not necessarily all) pairs of points.

One interpretation of the graph is that the points (also known as nodes or vertices) represent persons, and the connecting lines (also known as edges or links) indicate a relationship between two persons. For example, we might interpret the graph shown as indicating that B knows A, G, E, D, and H; that D only knows B; that E knows B, C, F, J, and I; and so on. There is no line between B and J, which indicates that B does not know J. However, if we imagine this to be the graph of a group of friends, we see that B can be introduced to J through H or through E.

In a more sophisticated analysis, we can quantify both the level of interaction and the direction of interaction. For example, if A and B see each other on

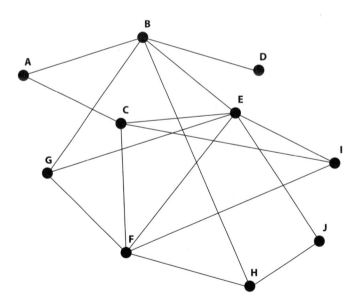

Figure 7. A Social Network

a daily basis, but A and C only meet once a month, we might give greater weight to the line between A and B than to the line between A and C: this produces a *weighted graph*. If this graph represents a hierarchical organization, such as a business or a criminal enterprise, A might give orders to B but B would never give orders to A; thus there might be a line *from* A *to* B, but not the other way around, which we could represent using a "one-way" arrow: this produces a *directed graph* or *digraph*. Or we could combine both weights and directions to produce a *weighted digraph*. There are also *hypergraphs* which associate three or more objects in a single connection: for example, if A communicated to B via email, but to C via phone, then A-email-B would be a connected triple, while A-phone-C would be another connected triple. In this case, two devices would never be connected, and the connection between two persons would always include at least one device. For simplicity, we'll focus on unweighted, undirected graphs, with the observation that everything that follows can be generalized to weighted graphs, digraphs, and even hypergraphs.

It is almost impossible to look at a graph like the one in figure 7 without wondering if there is any significance to the fact that some of the points seem better connected than others. For example, E is connected to six other points. In contrast, D is connected to just one other. Early attempts to use graphs to analyze the structure of social organizations focused on *degree centrality*: the number of lines coming from a given point. Thus E has degree 6, while D has degree 1. Intuitively, a point with higher degree is "better connected" than a point with lower degree.

There are other ways to gauge the importance of a point. The most obvious emerges as follows. Imagine someone (say A) has some information to be passed on to everyone in the group. For example, A might have a product to sell and believe that *someone* in the group would want to purchase it. This information can be passed to B and C. B and C could in turn pass on word of the product's availability to everyone they know (namely D, E, G, and H for B, and E, I, and F for C). In this way the information can be propagated throughout the group.

Intuitively, the fewer intermediaries the message must pass through, the faster the information propagates. In A's case, we have:

1. Two points, B and C, which are one connection away.
2. Six points, D, E, G, H, F, and I, which are two connections away.
3. One point, J, which is three connections away.

In some sense, the *total* distance from A to *everyone* in the group is then $2 \times 1 + 6 \times 2 + 1 \times 3 = 17$. From this, we can conclude that the mean distance from A to any other person in the group is $17/9 \approx 1.89$.

In contrast, consider someone like E. For E, there are:

1. Seven points, B, C, F, G, H, J, and I, which are one connection away.
2. Two points, A and D, which are two connections away.

Thus the total distance from E to everyone else is $7 \times 1 + 2 \times 2 = 11$ and the mean distance $11/9 \approx 1.22$. In some sense, more of the graph is closer to E than to A.

Imagine that traversing each connection takes some fixed amount of time. Then the reciprocal of this distance is a measure of how rapidly a "shout" from a point can propagate through the group. Social network analysts define the *nearness centrality* of a point as the reciprocal of the mean distance. Since the sum of the distances from E to the 9 remaining points of the graph is 11, the nearness centrality of E will be $9/11 \approx 0.82$. Similarly, the nearness centrality of A will be $9/17 \approx 0.5294$.

We might look at this measure in another sense. Consider a graph where everyone is directly connected to a central point P; for obvious reasons, such a configuration is called a *star*. In such a graph, every other point is 1 unit away from P, so the sum of distances and the number of other points will be the same; it follows that the nearness centrality of P will be 1. Points with a nearness centrality close to 1 share connectivity similar to the central point of a star.

There is a third common measure of centrality. Suppose two persons in the network wish to communicate with each other. Generally speaking, there are any number of paths between two points, though clearly the paths with the fewest connections are the most efficient; these paths are called *geodesics* and are equivalent to straight lines. However, while in plane geometry one and only one straight line exists between two points, it's possible for there to be two or more geodesics between two points. For example, between B and J there are two geodesics: B to E to J, and B to H to J. Both paths cross two connections, so both are "shortest."

Now suppose B and J wish to communicate using a geodesic. E is on one of the two geodesics between B and J. If we assume that the two geodesics are equally likely to be used, this means that E will handle half the communications between B and J. More generally, an individual on many geodesics is in a position to handle a great deal of information. The *betweenness centrality* measures this: the higher a person's betweenness centrality, the more geodesics he or she is on. If we assume that any two persons in the network are equally likely to communicate with each other, and are equally likely to use any geodesic between them, then the betweenness centrality of a point corresponds to the probability that the message will pass through the given point.

There are other measures of centrality, but we will focus on degree, closeness, and betweenness, as they are both the simplest and the most commonly used measures in social network analysis. The *key player problem* seeks to identify the most important individuals in a social organization. The centrality measures offer some insight: an individual with a high degree of centrality can connect directly with many others; an individual with a high closeness centrality can connect quickly with the entire network; and an individual with a high betweenness centrality is privy to much of the network's internal communication.

The Peril of Social Network Analysis

Social network analysis gained prominence following the attacks of September 11, 2001. After the attack, Valdis Krebs, the founder and chief scientist of the consulting firm Orgnet, used publicly available information to construct a graph, where the points represented the terrorists and the lines, close associations between them (those who had lived together or had attended school together). Krebs then studied the graph using the tools of social network analysis and found a number of remarkable features.

First, the four highest degree centrality, closeness centrality, and betweenness centrality scores were held by the same people (in slightly differing orders). Of particular interest is that these four people represented three of the four flights. In a "what if" scenario, we might wonder what would have happened if their high ranking drew the attention of law enforcement. It is also interesting to note that the fourth flight, which crashed into a Pennsylvania field, did *not* have any representatives in the "top four," and we might also wonder if this fact (which corresponds to a lower degree of integration into the group) and their failure are related.

Another well-studied group is the Jemaah Islamiyah terrorist cell that bombed a Bali nightclub on October 12, 2002. Stuart Koschade, of the Queensland University of Technology in Australia, considered a graph where the points of the graph represented the terrorists and the lines, significant contacts between them (either because of the duration or frequency of the contact). Koschade found that the field commander had by far the highest degree, closeness, and betweenness centrality, with the person in charge of logistical support for the operation a somewhat distant second. It is particularly interesting that the Bali bombers consisted of two teams whose only connection was the field commander. Again, our "what if" scenario causes us to wonder what would have happened if he had been taken into custody prior to the attacks.

This leads to a disturbing possibility. Suppose we could use social network analysis to identify key figures in a terrorist or criminal organization. We might be tempted to eliminate these figures to disrupt the organization. Thus on September 2, 2010, a drone strike targeted a convoy believed to be carrying a Taliban deputy governor and his bodyguards; this identification was made on the basis of signals intelligence and social network analysis. However, Afghan officials insisted that the convoy was a group of election workers seeking to promote democracy, and that the deputy governor was alive and well in nearby Pakistan. The problem is that social network analysis is based on patterns of interaction, not on individual characteristics: thus someone may have the same interaction pattern as an active member of a conspiracy *without* actually being involved in the conspiracy.

In particular, social network analysis may miss crucial features that distinguish a criminal or terrorist network from a more innocuous group. For example, social groups tend to be gregarious: your friends tend to know each other. A group of k points, where every point is connected to every other point, is said to form a k-clique. In figure 7, the points G, E, and F form a 3-clique, while C, E, F and I form a 4-clique.

Granovetter's rule is the observation that strong ties (where individuals are closely connected) tend to produce 3-cliques. For example, suppose A calls B and C frequently; this might be construed as there being a strong tie between A and B, and between A and C. We would expect, following Granovetter's rule, for a strong tie to exist between B and C: in other words, we would expect B and C to call each other frequently.

But what if they did not? For example, successful terrorist groups are organized to minimize the interaction between members, lest the betrayal of one lead to the betrayal of all. If we saw a violation of Granovetter's rule, say A contacting B, and A contacting C, but B never contacting C, we might suspect that these three form some sort of terrorist cell. At the same time, there are perfectly innocuous situations that lead to a violation of Granovetter's rule: A might be a parent, and B and C siblings who are not on speaking terms. The pattern of interactions alone cannot distinguish between terrorist cell and dysfunctional family.

Arguably, the internationally recognized doctrine of exigent circumstances allows governments to bypass certain legal safeguards when the situation requires an immediate response, and the "fog of war" will invariably lead to attacks on the wrong target based on flawed intelligence. However, in a civilian setting, we can and should be much more cautious, and avoid taking precipitous actions on insufficient information. The law holds that no single piece of evidence should be viewed as conclusive; if we maintain this viewpoint, we can shift the threat of social network analysis to 4th Amendment violations.

A key issue concerns telecommunication privacy. The *content* of interpersonal communication is protected in two ways. First, it is protected by the 4th Amendment's freedom from unreasonable search: a warrant must be obtained. More effectively, it can be protected by talking in code: even a legal wiretap could not determine the meaning of a phrase like "Don't forget to buy cream."

On the other hand, the *existence* of a communication between suspected conspirators is less well protected. If the communication is facilitated by a third party (such as the phone company or Internet service provider), the bar to obtaining the records of the communication is much lower. Moreover, the laws of physics make it difficult to conceal the existence of the communication.

In May 2006, allegations arose that the National Security Agency (NSA) was in the process of acquiring an enormous database of calls and callers. Such a database could potentially be used to ferret out suspicious communication patterns, such as those that violate Granovetter's rule. This raises a new (and as yet unsolved) constitutional question: at what point does one's participation in a network with "suspicious" interaction patterns constitute grounds for a warrant?

The problem is particularly acute because terrorists and criminals also interact with law-abiding members of the public. We need a way to distinguish between people who interact with terrorists socially, because they live in the same neighborhood or shop at the same grocery stores, and people who interact with them because they believe in the use of violence to attain political goals.

Terrorist and criminal networks do not publish organizational charts, nor do they identify their members. Thus, while a terrorist network might in fact be organized as shown in figure 7, law enforcement might have a very different view of the organization: edges and even nodes might be missing. For this reason, terrorist and criminal organizations are sometimes referred to as *dark networks*, since much of the network structure must be considered unknown.

Nevertheless, law enforcement might be able to glean some information from the known members of the network. Christopher Rhodes, at the Imperial College in London, considered this problem and took as his starting point the analysis of the (defunct) Greek terrorist group November 17. Imagine we are confronted with a dark network, and know some but not all of its members. With infinite time, money, and resources, and a compliant court willing to issue warrants on demand, we could watch every known member of the group 24 hours a day, and have some chance of identifying all members. But in practice, this is impossible.

Consider two members of such a network. These members are either connected, indicating some sort of substantive connection between them, or not. In general, we might consider the probability that a connection exists; this is measured by the *density* of the network, defined as the number of actual connections divided by the number of connections that could exist. Dark networks, at least the successful ones, tend to have low densities: this ensures that if any individual is compromised, he or she cannot reveal much information about the rest of the network. For example, the September 11 terrorist network had a density of about 16%. This means that two randomly chosen members of the network had a 16% chance of ever having made contact with each other. It follows that if we simply observe known members of the conspiracy in the hopes of finding unknown members, our yield will be low.

However, as with our data mining example, we can revise our estimate of the probability if we have some additional information; this allows us to focus our surveillance on individuals most likely to have contact with other members of the conspiracy. Rhodes proposed using a Bayesian approach, based on a simple notion: associations are not formed at random, but are based on some common interest or need.

To proceed, we assume complete knowledge of a similar organization: for example, a terrorist group that had been dismantled by the authorities. Statistical analysis of this *gold network* can then be used to infer relationships in the dark network. Rhodes used a six-member subgroup of November 17,

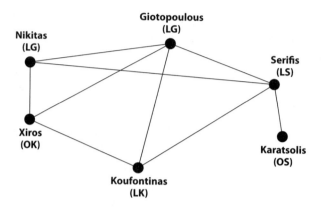

Figure 8. Gold Subnetwork of November 17

which might be viewed as a cell of the terrorist organization that law enforcement has identified. In this subgroup, shown in figure 8, members could be classified according to various attributes: they might be leaders (L) or operational individuals (O); they might belong to the first generation (G), or to factions (S) or (K) within the group. They might also have access to different resources: weapons, money, safe-houses, and so on, and be involved in subsidiary activities like human trafficking or bank robberies (not shown on the graph, though this information could be included in a hypergraph).

To begin with, 6 individuals can be connected in 15 ways, though in this particular network, only 9 connections existed. Thus gold network's density is 9/15 = 60%. We can take this as our prior probability that two individuals are connected.

There's a more subtle bit of information we can extract from the gold network: If 9 pairs of individuals are connected, and 15 pairs are possible, then there are 6 pairs of individuals that are *not* connected. The non-connections are important, and unique to the gold network. In particular, failure to observe a connection between individuals in a dark network merely means that we have not observed a connection. In contrast, since we assume complete knowledge of the gold network, the absence of a connection means that no connection exists.

Next, we observe that the six individuals in the gold network included four members classified as leaders (L). Four individuals can be connected in six ways, and if they are no more closely connected than the network in general, we would expect to see 60% of four—or between two and three connections

among them. Intuitively, we would expect leaders to be more closely connected; this is what we observe, with five connections between the leaders
and one non-connection.

How shall we interpret this observation? Consider two randomly chosen
individuals in the organization. The prior probability that a connection exists between them is 60%. But what if we know that the two individuals involved are leaders? In this case, we should be able to revise our estimate of
the probability they are connected using Bayes's theorem and the likelihood ratio.

Let C be the event that a connection exists between two individuals, and L
be the event that two individuals are leaders. Bayes's theorem gives us

$$\frac{P(C|L)}{P(\text{not } C|L)} = \frac{P(L|C)}{P(L|\text{not } C)} \frac{P(C)}{P(\text{not } C)}.$$

Thus our likelihood ratio will be the probability that a connection joins two
leaders, divided by the probability that a non-connection "joins" two leaders.
Since 5 of the 9 connections are between leaders, the first probability is 5/9;
since 1 of the 6 non-connections are between leaders, the probability of the
latter is 1/6; and our likelihood ratio is (5/9) / (1/6) ≈ 3.33. Leaders are a little
more than three times as likely to be connected as randomly chosen members of the group.

The likelihood ratio exists even if we don't know the identities of both individuals. This suggests the following relatively benign use of this information. Suppose law enforcement wishes to place an informer in the group. The
most effective strategy is to have the group recruit someone from law enforcement. A study of the likelihood ratios will identify the characteristics of
a person most likely to be contacted by the group, and law enforcement could
themselves recruit an informant meeting those characteristics.

Another use, somewhat less benign, is the following. Suppose we suspect
there are other leaders beyond those we already know. We could devote law
enforcement resources to watching non-leaders, in the hopes that they would
eventually contact leaders of the group. But the likelihood ratio allows us to
identify the characteristics of persons most likely to contact a higher level of
leadership; law enforcement could then obtain a warrant authorizing detailed
surveillance of these individuals. As with data mining, we must ask the question: At what point does the likelihood ratio become large enough to constitute probable cause?

The Promise of Social Network Analysis

At the same time it threatens 4th Amendment freedoms, social network analysis offers considerable promise and support for 8th Amendment rights. In particular, centrality measures offer a way to quantify an individual's involvement in a group. While this might not prevent a defendant from being convicted, it could be used to compare his or her sentence to those of similarly placed individuals.

To that end, we might ask if there is any correlation between an individual's centrality measure and his or her sentence. Research on this topic is still in its infancy; however, preliminary results suggest that the correlation does exist.

For example, in 1958 the Tennessee Valley Authority solicited bids for the construction of the Colbert Steam Plant, and received a number of identical or nearly identical bids from contractors. The Department of Justice investigated, and in 1960 handed down a number of indictments for price fixing, which involved collusion between managers of a number of companies manufacturing electrical generation equipment. Unusually for the time, the prosecution declined to accept *nolo contendere* ("no contest") pleas in exchange for token fines and immunity from further legal consequences. Instead, they chose to pursue criminal trials against the defendants, and won a number of convictions and substantial fines.

In 1993, Wayne Baker of the University of Chicago and Robert Faulkner of the University of Massachusetts analyzed the conspiracy using the tools of social network analysis. If the points in the network represent the managers involved in the conspiracy, and the lines indicate collusion (as reported by the managers themselves during their trials), then there is a clear correlation between degree centrality and the likelihood of conviction.

A 2013 study of the sentencing of members of a Canadian drug ring known as the Caviar Network showed a similar correlation between the severity of a sentence and a defendant's degree centrality; this study also showed a correlation with closeness centrality. While the centrality measures played no role in the actual sentencing, it suggests that they can play a role in assessing the proportionality of a sentence after-the-fact.

Social network analysis has great promise when dealing with *vicarious liability*, a measure of the culpability of one person for the crime committed by another. In general, the legal system discourages the concept of vicarious liability: you are only responsible for your own actions. But under current U.S. law, *everyone* involved in a drug distribution ring, regardless of their level of involvement, could be subjected to the same penalty. Here the need to evaluate a person's level of involvement is serious: in the United States, about 28,000 persons each year are convicted of violating federal drug trafficking laws.

The challenge is that a conspiracy to make and distribute illegal drugs extends through time and space. Some participants might never meet each other; there may never be a point in time where all participants are in the same vicinity; and in fact, it is entirely possible for some members of the conspiracy to be unaware of its nature.

The courts might reconsider the case of Theresa McIntyre Smith in light of social network analysis. Her connection to the network was primarily through Mercer; as a result, she would have a very low degree, closeness, and betweenness centrality. We might then compare her sentence to those of defendants with similar centrality measures. If we find too great a disparity between them, this might be used as grounds for declaring the sentence disproportionate to the offense.

Three Strikes for Three Strikes

> Objective criteria clearly establish that a mandatory life sentence for
> defrauding persons of about $230 crosses any rationally drawn line
> separating punishment that lawfully may be imposed from that which
> is proscribed by the Eighth Amendment.
> —JUSTICE LEWIS POWELL, *RUMMEL V. ESTELLE* (1980)

William James Rummel was sentenced to life in prison in a Texas court on
April 26, 1973, for violation of Texas's repeat offender laws. Simply put, Rummel had broken the law too many times. His crimes: first, in 1964, he fraudulently used a credit card in the amount of $80, for which he was given a three-year prison sentence. Second, in 1969 he passed a forged check for $28.36, for which he was given a four-year sentence. Third, in 1973, he received $120.75 to repair an air conditioner, and did not perform the work.

Rummel contested his sentence, arguing that a life sentence for his crimes constituted cruel and unusual punishment within the meaning of the 8th Amendment. Justice Powell noted, "It is difficult to imagine felonies that pose less danger to the peace and good order of a civilized society than the three crimes committed by [Rummel]" (445 U.S. 263, p. 294). In fact, following Rummel's conviction and sentencing, the state of Texas reclassified Rummel's third crime, theft by false pretexts, as a misdemeanor, yet another point in favor of Rummel's viewpoint. Other members of the U.S. Supreme Court agreed that the sentence was disproportionate. Nevertheless, in a 5 to 4 decision, the Supreme Court upheld the constitutionality of Rummel's sentence.

Rummel ran afoul of an approach to punishment known as a *repeat offender law,* which imposes extended or even life sentences upon conviction for a specified number of felonies. Such laws have a long history in the United

States; New York State was the first to pass one, in 1796, mandating life sentences for persons convicted of a third felony. However, most states abolished these laws, and by the 1970s, only three states retained them: Washington, West Virginia, and Texas.

The California legislature considered a repeat offender law in 1993, but rejected it: the cost of keeping a prisoner incarcerated is high (currently around $35,000 per prisoner per year), and the wisdom of expending this much money for nonviolent offenders like Rummel was dubious. A simple-minded strategy that imposed extended sentences based solely on the number of convictions, as opposed to their barbarity, type, or frequency, would be financial folly.

Three strikes might have gone into the dustbin of discarded theories of punishment, but shortly after the California Assembly rejected a repeat offender bill, 11-year-old Polly Klaas was abducted from her home, raped, and murdered by Richard Allen Davis. Davis was on parole for a previous kidnapping, and the specter of a convicted felon being free to commit heinous crimes drew national attention to the problem of recidivism.

Opposition to the three strikes law vanished. In his State of the Union address, then-President Bill Clinton called for a federal three strikes law. In California, supporters gathered enough signatures to put Proposition 184, imposing life sentences on repeat offenders, on the November 1994 ballot. California legislators, not wanting to appear laggard, passed their own repeat offender law in March 1994; voters passed Proposition 184 in November; and within two years, about half the states and the federal government had such laws.

As in Texas, California's simple-minded approach produced some extreme punishments for trivial crimes. Thus Leandro Andrade would receive two consecutive 25-years-to-life sentences (a total of 50 years before being eligible for parole) after stealing nine videotapes over a two-week time period in 1995; and Gary Ewing would receive a 25-years-to-life sentence for stealing three golf clubs from the pro shop at the El Segundo Golf Club in 2000.

In fairness to defenders of draconian penalties for repeat offenders, the previous crimes of Andrade and Ewing posed rather more danger to the "peace and good order of a civilized society" than Rummel's. Andrade had committed a burglary (theft of property from an unoccupied residence) and escaped from prison. Ewing's crimes were even more serious, including a string of robberies (theft of property by force or threat of force). Thus in *Lockyer v. Andrade* (2003) and *Ewing v. California* (2003), the Court upheld the extended sentences handed down to Andrade and Ewing.

The Purpose of Punishment

The question we must confront is why we are punishing an offender. There are three primary goals of punishment: retribution, to make the perpetrator "pay" for their actions; incapacitation, to prevent the perpetrator from committing similar actions in the future; and deterrence, to discourage others from following in the perpetrator's footsteps.

Retribution has largely fallen out of favor in the criminal justice system, though vestiges remain in the guise of parking fines and speeding tickets, as well as the occasional fines handed down to corporations that break the law. Despite the rhetoric of "paying for your crimes," a prison sentence does not constitute retribution: the victims receive nothing whatsoever when an offender is incarcerated. Instead, prison sentences serve the purpose of incapacitation: the offender is prevented from committing another offense. Obviously, a person sentenced to life in prison poses little danger to the general public.

The problem is that this argument applies to bank robbers and jaywalkers alike. We might argue abstractions: jaywalking is not so serious a crime as bank robbery, so to punish jaywalking with a life sentence is excessive. However, the Supreme Court has upheld the state's authority to impose extreme punishments; if it is permissible to impose a life sentence for petty larceny, there is no a priori reason we couldn't impose a life sentence for jaywalking. The classification of jaywalking as a misdemeanor is not actually relevant: in *Duncan v. Louisiana*, the Supreme Court ruled that it is the severity of punishment, and not the classification, that determines the seriousness of the crime. In other words, if you could be sentenced to life in prison for jaywalking, then jaywalking must be considered a major crime.

There is another angle of approach: a three strikes law criminalizes repeat offending. In particular, the extended sentence is *not* for the crime that was committed. It is incorrect to say that Rummel was sentenced to life for refusing to refund $120.75, or that Andrade was given a sentence of at least 50 years for stealing videotapes of *Snow White*, or that Ewing was sentenced to 25 years in prison for stealing golf clubs. Rather, these offenses, together with their prior offenses, constituted a new offense. This is important from a constitutional perspective as well, since the 5th Amendment prohibits punishing a person twice for the same crime.

Game theory allows us to evaluate the two punishment strategies as follows. First, we can punish the third offense as prescribed by law: thus Rummel's third offense would have ordinarily drawn a sentence of 2 to 10 years in prison. Alternatively, we can punish it with an extended sentence, as prescribed by a three strikes law. We can then perform a cost-benefit analysis of the two strategies, to see which is the more reasonable, and determine which strategy is more cost effective: philosophical, ethical, and moral arguments are good, but few arguments are as persuasive as that of the pocketbook.

To begin with, there is one easily quantifiable cost of repeat offender laws: maintaining a prison costs money. Even if we spend as little as possible maintaining the prisoners themselves, there are other costs: the guards must be paid (and paid well, lest the temptation of bribery become too powerful); the building must be maintained (and maintained well, lest escape become too easy); and administrators must be in place to oversee the facilities. It costs between $30,000 and $40,000 each year to keep one person in prison. Assuming a normal lifespan and ignoring inflation, we might take $1,000,000 as the estimated cost of sentencing a repeat offender to life over and above the cost of imposing the legally mandated sentence for the specific offense. Does society reap at least $1,000,000 in benefits?

In order to answer this question, we must confront another, difficult question: What is the cost of a crime? It's tempting to calculate the cost based on the harm inflicted on society: after all, we do talk about a felon's "debt to society." Hence seat belt laws are justified by the fact that a person who gets into an accident wearing a seat belt is less likely to require expensive emergency room treatments.

This analysis is more difficult with property crimes. Aesop's fable of the miser and the thief illustrates a basic economic fact: the value of property comes from its use, not its possession. Society is not harmed if possession of an object transfers from its rightful owner to a thief. As with all cost-benefit analyses, we must ignore non-quantifiable abstractions like "law and order."

Rather than focusing on the effects on society, we should consider the actual cost accrued to the victim as the result of a crime. For property crimes, this computation is straightforward. Violent crimes are somewhat more

Table 14.   Costs of Crime (in 1993 dollars)

Crime	Tangible Losses	Quality of Life	Total
Murder	1,030,000	1,910,000	2,940,000
Rape	5,100	81,400	87,000
Assault	1,550	7,800	9,400
Robbery	2,300	5,700	8,000
Burglary	1,100	300	1,400
Larceny	370	0	370

complicated. There are the obvious costs of medical treatment, for both physical and psychological wounds. To these we can add productivity costs: work days lost due to recovery or posttraumatic stress, or even permanent disability resulting from a criminal act. However, what is the cost of being afraid to leave one's home?

One of the seminal studies on this topic appeared in January 1996; it was written by Ted Miller, of the National Public Services Research Institute; Mark Cohen, of Vanderbilt University; and Brian Wiersema, of the University of Maryland. To assess the intangible ("quality of life") costs of crime, they used a variety of sources, including punitive judgments by juries in civil suits against wrongdoers. We include some of their results in table 14.

First, consider Rummel, whose underlying crimes fall into the "larceny" category. If we assume that he would have continued this petty thievery for the rest of his life, interrupted by prison sentences, then we can estimate the value of the crimes prevented by a life sentence. In Rummel's case, we might suppose that each conviction for theft might have resulted in a 5-year sentence: thus he might have been sentenced six more times over the course of his life for six more counts of theft or check forgery.

The times he would have been convicted are only a fraction of the times he offended. Unfortunately, it's generally impossible to determine how many crimes a person committed before he or she is actually convicted of a crime. One methodology is to use surveys of convicted criminals. Based on such surveys conducted of inmates in selected counties in California, Texas, and Michigan, and another of inmates in the cities of Detroit, Michigan, and Washington, DC, Cohen and Alex Piquero, of the University of Maryland, obtained a set of *offense multipliers* for various crimes (provided in table 15).

Table 15 has some interesting features. For example, in Detroit and Washington, DC, the typical robber committed about 17.8 robberies for each actual

Table 15. Offense Multipliers

Crime	Detroit and Washington, DC	California, Texas, Michigan
Murder	1	1
Rape	11.6	4.2
Assault	11.6	4.2
Robbery	17.8	5.4
Burglary	23.0	16.7
Larceny	35.7	50.0

conviction. In contrast, California, Texas, and Michigan law enforcement more aggressively pursued robbers, who only committed an average of 5.4 offenses before being caught and convicted. On the other hand, law enforcement in Detroit and Washington more actively pursued larceny, with each perpetrator accomplishing "only" 35.7 acts before being caught, as opposed to 50 in counties in California, Texas, and Michigan.

If we use the figures for Texas, where Rummel was convicted, we might estimate that he would commit 50 acts of larceny for every actual conviction. That means Rummel's six future convictions would translate into some 300 additional acts of larceny. Using the $370 figure for the estimated victim cost of these acts (which, we note, is more than the *total* amount of all three of Rummel's prior offenses), we find that for $1,000,000, the state of Texas prevented $370 \times 300 = $111,000 in crime: a bad bargain by anyone's standards.

Andrade's case is similar. Again, assuming a 5-year sentence for each subsequent offense, Andrade might have been convicted of six more burglaries over the remainder of his life, while committing as many as 100, each with an average cost to the victim of around $1,400. Thus a $1,000,000 investment prevented $140,000 in crime.

Ewing's case is more difficult: though his third strike was for burglary, his prior offenses included violent crimes like robbery. If we assume all his future convictions would be for burglary, then his case is similar to Andrade's; if we assume that they would have been for robbery, the more costly crime, a $1,000,000 investment prevented $256,000 in crime.

Markov Chains and Escalation

The real fear is that an offender escalates and commits a more serious crime, such as the rape and murder of Polly Klaas. Some insight into how often this

occurs can be gleaned from table 16, based on a study of 272,111 felons released in 1994. The study followed this cohort for three years after their release, to evaluate how likely they were to return to prison for a new offense.
The columns show the most serious offense they had committed prior to
their release, as well as the total number released in 1994; the rows show the
offense that led to their re-arrest.

One way to use this information is to estimate the number of crimes committed by released felons within three years of their release. Consider murder,
for example. According to table 16, 1.2% of murderers go on to commit another murder within three years of their release: thus the 4,443 murderers
released in 1994 would be responsible for about 53 murders over the next
three years. However, some murders are committed by those convicted of
other felonies. The largest contribution comes from those convicted of assault: not only is this group the largest, at 17,708, but they are most likely to
escalate to murder: within three years of their release, this group would be
responsible for about 283 murders, or more than five times as many as those
actually convicted of murder.

Table 16 provides an example of a *stochastic matrix*, and we can use it to
create a *Markov chain*, named after the Russian mathematician Andrey Markov (1856–1922). Markov's work should be considered by anyone who questions the value of abstract research: it began by an examination of how conso-
nants and vowels distributed themselves in a work of poetry, but led to one of
the most powerful tools in applied mathematics.

A Markov chain can be used to describe a system that can be in any number of discrete states. Imagine that felons are sent to islands according to
their crime: thus there is the island of murderers, the island of rapists, the
island of assaulters, and so on. After their time is served, the felons are re-

Table 16.   Re-offense Rates (%)

	Murder	Rape	Assault	Robbery	Burglary	Larceny	Other
Released in 1994	4,443	3,138	17,708	26,862	41,257	26,259	132,176
Murder	1.2	0.7	1.6	1.1	0.7	0.6	0.8
Rape	0	2.5	1.0	1.2	0.8	0.5	0.4
Assault	11.9	8.7	22.0	15.1	13.8	14.4	12.4
Robbery	3.4	3.9	6.1	13.4	5.9	7.3	5.0
Burglary	2.0	4.4	7.7	8.7	23.4	13.9	6.0
Larceny	4.1	6.2	10.6	16.5	23.0	33.9	12.2
Other	77.4	73.6	51.0	44.0	31.0	29.4	63.2

turned to the mainland (the population at large). The recidivism rates correspond to the likelihood that a felon on an island will, after release, return to one of the islands in the prison archipelago. For example, a person on the island of murderers has a 1.2% chance of returning there; a 0% chance of going to the island of rapists; an 11.9% chance of going to the island of assaulters; and so on.

For illustrative purposes, suppose we are considering a "one strike" rule: one conviction for murder, rape, assault, robbery, burglary, or larceny receives a life sentence. While this rule seems draconian, it is in effect the three strikes rule when we consider only the last conviction. Also for for illustrative purposes, assume that a felon will have the *opportunity* to commit six additional crimes if not given a life sentence.

Note that the opportunity to commit a crime does not mean that a crime is actually committed. For example, a murderer has a 1.2% of committing another murder within three years after release—which means he or she has a 98.8% chance of *not* committing another murder within three years, and in fact a convicted murderer has a 77.4% chance of not committing *any* serious felony within three years after release.

The stochastic matrix allows us to make predictions of how someone will travel through the islands. For example, we can predict that, under our assumptions, a typical murderer will be arrested for 0.04758 murders after release, 0.02296 rapes, 0.68573 assaults, 0.27074 robberies, 0.33379 burglaries, and 0.69146 larcenies. Of course, one can't be arrested for a fraction of a murder; these numbers mean that 4.758% of those who start on the island of murderers will return to it at some point; 2.296% will spend at least some time on the island of rapists; and so on.

As above, we can apply the offense multipliers: thus the released murderer will likely be responsible for 0.04758 murders, 0.09644 rapes, 2.880 assaults, 1.4620 robberies, 5.5743 burglaries, and 34.573 larcenies. Tallying the costs of these crimes, we find that a life sentence for murder would prevent about $207,649 in future crimes committed by the murderer. We could use this to debate the wisdom of spending $1,000,000 to save $207,649, but a better use might be the following. Suppose we agree that a life sentence for murder is reasonable. Then we should regard life sentences as reasonable for *any* crime with a higher future cost. Thus we would also impose life sentences for those who commit assault ($235,558 in future crime costs), robbery ($222,910 in future crime costs), and burglary ($209,632 in future crime costs).

On the other hand, suppose we deem a life sentence for murder to be not worthwhile. Then we should likewise reject life sentences for any crime with a *lower* future cost, that is, if murder does not receive a life sentence, neither should rape ($201,267) or larceny ($206,933).

This is a very simple model. One of the key assumptions is that the probability of switching from one offense type to another remains constant. This appears to be true in the short run, but research indicates that over time, criminals become specialists; the trend is particularly noteworthy for sex offenders and those who commit fraud. In other words, the prior history of offenses makes a difference.

This makes the problem more difficult, but not necessarily more complicated: instead of an island of murderers, there is an island of murderers-robbers-assaulters (indicating that the person first committed an assault, then a robbery, then a murder) and, since order matters, a nearby island of robbers-assaulters-murderers (indicating a slightly different sequence of criminal activity), and so on. Our stochastic matrix would be larger, and the data required to construct it more extensive, but the essential approach would be unchanged.

If we are willing to embrace more complexity, we can use a *hidden Markov model.* As with a basic Markov chain, we assume that there is a stochastic matrix that describes the transition between different states. The key difference is that the states are not directly observable; we can only observe their consequences. For example, suppose our states represent an offender's mind-set. One state might be "murderous," and this could lead to an actual murder. On the other hand, the same mental state might lead to a different crime, say "assault." Likewise, a "larcenous" mind-set could conceivably lead to a murder, though it is more likely to lead to a larceny, burglar, or robbery. The crimes we can observe are functions of the states we cannot.

Setting up a hidden Markov model is much more difficult, since we need to construct two matrices: the stochastic matrix, as in the standard Markov chain; and a second matrix that models the probability that a hidden state produces an observable effect. The techniques for doing this have only become computationally feasible in the past decade, so we are at the very beginning of what can be done with them. However, early results seem promising.

For example, in 2006, Philip A. Schrodt of the University of Kansas constructed a hidden Markov model for the situation in the Balkans between January 1, 1991, and January 1, 1999. During this time period, the communist nation of Yugoslavia descended into civil war and, after much bloodshed and atrocity, became the modern countries of Bosnia, Croatia, Macedonia, Serbia, and Kosovo. Schrodt used the level of conflict as the hidden states, and the number of violent incidents as the observed events, and produced a model that used one week's level of violence to forecast, with some reliability, the next week's level of violence.

A similar approach was taken by Francesco Bartolucci, Fulvia Penoni, and Brian Francis, three European researchers. In 2007, they published their results, a study of a cohort of offenders born in 1953, and found the approach provided a good match between predicted and observed re-offense behavior. They suggested that the results could be particularly useful to policymakers when crafting sentencing policy.

One possibility is to use predicted future behavior as a basis for sentencing. In 1993, George Allen became governor of Virginia, on a platform that included eliminating parole and increasing penalties for violent offenders. The next year, the General Assembly created the Virginia Criminal Sentencing Commission (VCSC) to study the feasibility of placing 25% of nonviolent felons into alternative (non-incarceration) arrangements. Obviously, the 25% should be those who pose the least danger to society, so the real task is determining the risk posed to society by a given felon.

As a first step, the VCSC studied a sample of roughly 1,500 nonviolent felons incarcerated between July 1, 1991, and December 31, 1992. By tracking this cohort, they determined re-offense rates; a sophisticated statistical analysis led to a worksheet that used 11 factors to determine the likelihood of recidivism and to offer a sentencing recommendation. The use of this risk assessment worksheet is now mandatory in sentencing nonviolent felons. While judges are free to impose harsher or more lenient punishments, about 81% of sentences follow the worksheet recommendations.

In theory, such a model can be extremely useful. However, a mathematical model is a power tool, and if we are not careful to read the instructions and follow all the safety guidelines, we can do a great deal of harm with it. In Virginia's case, the model was based on a study of the recidivism rates of released offenders. Ironically, this means the most appropriate use of such a worksheet is to guide parole decisions—the very system abolished by Governor Allen in 1995!

The problem is that the worksheet is based on the study of released offenders who have already served their sentences. Prisons are often accused of being little more than places where felons can hone the skills of their trade; following this logic, we might question the use of information about released offenders to guide decisions about first-time offenders.

Instead, suppose the parole system had remained in place. Under the existing system, very few offenders receive the maximum penalty prescribed by law. But suppose instead that each offender received a sentence with a specified parole ineligibility period: for example, a sentence of eight years with two years to pass before parole is even considered. At the end of two years, their case would be reviewed using the risk assessment worksheet. Those who posed little danger to society could be released, while the remainder would serve the rest of their sentence. This approach is equivalent to a system without parole that sentences some offenders to 2 years and others to 8 years. The main difference is that instead of requiring the judge to make a sentencing decision at the instant of conviction, a more rational sentence can be imposed, based on inmate behavior during their incarceration.

The Deterrence Effect

The preceding analysis suggests that the cost of imposing life sentences on repeat offenders is significantly greater than the cost of the crimes they commit were they not incarcerated. Thus, three strikes laws cannot be justified on the basis of incapacitation.

However, there is another rationale for their existence: deterrence. Obviously, a lengthy prison sentence will deter someone from committing a crime that leads to it, so the crime rate will fall. And it is self-evident that a criminal facing a lengthy prison sentence might choose to murder witnesses that could lead to a conviction, so the murder rate will rise.

One should always be wary of sentences that include phrases like "obviously" or "it is self-evident," since these phrases indicate a conclusion derived from logic. The problem is that a logical conclusion is only as good as its premises. This is sufficient for theoretical mathematics, but if we want to apply mathematics to the real world, we must ensure that our premises are reasonable, and the best way we can do that is to base them on quantifiable data.

One of the facts presented in support of the three strikes law is a simple one: the violent crime rate in California fell by 43% between 1994 and 1999. This is an impressive decline, and supporters attribute it to the three strikes law; indeed, the California Attorney General's office released a report in 1998 that claimed the three strikes law had resulted in 4,000 fewer murders and 800,000 fewer criminal victimizations. And if the three strikes law were the only change in society to occur between 1994 and 1999, we might agree.

But crime fell nationwide by about the same amount, even though only about half the states had repeat offender laws. Moreover, the 1990s witnessed an improving economy, the rise of community policing, the decline of crack cocaine, and the coming of age of the first generation after the legalization of abortion in 1973, all of which have been touted as reasons for a fall in the crime rate.

To disentangle the effect of the three strikes laws, we might compare the crime rates in different states: those that enacted such laws and those that did not. Moreover, we would want to make a comparison between states as similar as possible: for example, it would be disingenuous to compare California, a heterogeneous, highly populous state that espouses the death penalty and a three strikes law, with Minnesota, a much more homogeneous, much less populous state with neither the death penalty nor a repeat offender law. On the other hand, it might be more reasonable to compare California with New York, which has the death penalty but no three strikes laws, and Texas, which has both, though the Texas law had been in force for many years. The results are shown in table 17.

This pattern holds nationally: murders and violent crimes declined nationwide between 1994 and 1999, but in general declined *more* in states *without* three strikes laws. A statistical analysis shows that this difference is statistically significant, so it was not likely to be caused by purely random variations: some non-random factor caused the murder rate to fall by less in California than in New York. The obvious culprit is the three strikes law.

Table 17. Decline in Murders in Three States

	California	New York	Texas
1994 murders	3,703	2,016	2,022
1999 murders	2,005	903	1,217
Decline	46%	55%	40%
1994 violent crime	318,395	175,433	129,838
1999 violent crime	207,879	107,147	112,306
Decline	34.7%	38.9%	13.5%
2010 violent crime	164,133	75,977	113,231
Decline 1994–2010	48.4%	56.7%	12.8%

But again, we should be wary of any sentence that includes the phrase "obvious." The three strikes law is not the only difference between the two: as the old joke goes, New York is 3,000 miles from the ocean. A more sophisticated statistical analysis would take into account as many factors as possible. This can be done, and in 2001, Thomas Marvell of Justec Research, and Carlisle Moody of William and Mary College, both located in Williamsburg, Virginia, concluded that the three strikes laws were in fact associated with an increase in the murder rate.

The approach used by Marvell and Moody is essentially the same as that used to construct the seats-votes curve, though rather than being based on a single variable (the fraction of the popular vote won by a party's candidates), many variables can be incorporated. A very simple illustration of this approach is the following: If the murder rate in California fell by as much as it did in New York, there would have only been about 1,666 murders in California in 1999; instead, there were 2005—339 more than expected. These extra murders must be attributed to whatever differences exist between California and New York.

Marvel and Moody used a much more sophisticated approach to derive a formula that allows us to do by mathematics what we could never do with a real society: change just one thing and observe the effects. In this case, we can include the factor corresponding to a three strikes law, or omit it. If we include it, we find a higher murder rate than if we do not; Marvell and Moody concluded that a three strikes law raised the murder rate by 10–12% in the short run and by 23–29% in the long run. This rise has been masked by a long-term decline in the murder rate, but we can see it in the comparison of the California and New York numbers above: *something* caused about 339 more murders in California than we would expect.

An alternative approach to evaluating the effects of the three strikes law was taken in 2007 by Eric Helland of Claremont-McKenna College in California, and Alex Tabarrok, of George Mason University in Virginia. Rather than trying to compare crime rates before and after three strikes legislation passed, or comparing the crime rates in states with and without three strikes legislation, they considered the effect on the individual offenders: Does a person with two strikes offend at the same rate as a person with one strike? This approach has the advantage of removing any variability caused by different approaches to law enforcement, and can focus on what supporters of three strikes claim is its best feature: it keeps repeat offenders from re-offending.

Helland and Tabarrok found a statistically significant difference in arrest rates between those with one strike and those with two strikes. This suggests that three strikes *does* deter—at least, it deters those who are at risk of having a third strike. This deterrent effect may reduce the number of crimes committed by as many as 31,000 each year. However, as they point out, this does not provide any insight into the general deterrent effect. Moreover, they note that the real question is whether the deterrent effect can be obtained more efficiently by other means.

In particular, the three strikes law adds about $280 million each year to the cost of running California prisons. This amount could pay for perhaps 2,000 more police officers. Thus we might ask: Would 2,000 more police officers be able to prevent more than 31,000 crimes annually, either through their presence in the community, or through whatever deterrent effect increased crime-solving resources would produce?

Striking Out: Victory by Numbers

Motivated by the Polly Klaas case, California lawmakers and voters passed the most draconian sentencing law in the country in 1994; other states followed with similar legislation. Three strikes survived questionable sentencing decisions like a 50-year sentence for stealing videotapes or a 25-to-life sentence for stealing golf clubs. But arguments based on abstractions like fairness and justice failed to turn voters against the three strikes law. What ultimately worked was an argument based on mathematics, combined with a palatable alternative.

The rationale for three strikes is that a repeat offender will continue to commit additional crimes after release. But the cost of imposing a life sentence for those who commit nonviolent and property crimes far outweighs the cost of the crimes they would commit. Moreover, there is a statistically significant difference in the violent crime rate between states with three strikes laws and states without: namely, states without three strikes laws have lower violent crime and murder rates. Finally, while there does seem to be a deterrent effect on those who have a second strike, the magnitude of this effect must be weighed not only against the extra $280 million cost incurred by administering life sentences to all third strikers, but against alternative measures of crime control that could be bought for the same amount.

Thus in 2012, California voters considered Proposition 36, which modified the three strikes law as follows. First, a life sentence could only be imposed if the third strike was a violent felony. Second, those who had been sentenced for nonviolent felonies could appeal their sentences. The nonpartisan Legislative Analysts Office estimated that passing the measure would save at least $90 million a year. The measure passed, with nearly 70% of the voters supporting it.

We may take Proposition 36 as a blueprint for future action. For 18 years, activists had opposed three strikes on philosophical grounds like justice, fairness, and mercy. And for 18 years, three strikes persisted. The cold equations, combined with a palatable alternative, succeeded where philosophical arguments failed.

The Price of Punishment

The defendant of wealth and position never goes to the electric chair or the gallows. —LEWIS E. LAWES, WARDEN OF SING SING, 1920–1941

Around 12:30 am on September 7, 1984, Maria Marshall was shot and killed in an apparent robbery on the Garden State Parkway in New Jersey. But the police investigation revealed some disturbing factors. Maria's husband, Robert Marshall, was an insurance agent, who over the past few months had been increasing the value of the life insurance policy on his wife, while falling behind on payments to his own policy; at the time of her death, Maria was insured for about $1.4 million. Telephone records showed Marshall had contacted Billy Wayne McKinnon, a former sheriff's officer from Louisiana, and when police interviewed McKinnon, he turned state's evidence and explained that Robert Marshall had hired him to kill Maria. McKinnon would eventually be sentenced to five years in prison and a fine of $5,000, part of the plea bargain he arranged for his testimony. Marshall would be found guilty and sentenced to death.

Marshall requested a *proportionality review* to determine the appropriateness of the sentence of death. While the state granted his request, there was one problem: there was no consensus on how to conduct such a review. On July 29, 1988, the New Jersey Supreme Court appointed David Baldus Special Master, tasking him with designing a proportionality review process.

Baldus analyzed more than 2,000 homicide trials in New Jersey and created a mathematical equation that predicted, with some degree of accuracy, the probability that a defendant would be sentenced to death. Such a model can be useful in assessing a state's system of imposing the death penalty. However, because it returns a probability, it tells us nothing about any single case. For example, on May 5, 1992, Donald Loftin murdered Gary Marsh by shooting him in the course of a robbery. Baldus's mathematical models computed that the probability that Loftin would receive a death sentence was 14%. But just as we would probably not conclude anything amiss if we observed a coin land heads three times in a row, we would also be unperturbed by the fact that Loftin received the death penalty: both events are unlikely, though not so unlikely that we would be inclined to attribute their occurrence to anything other than the luck of the draw.

Baldus recommended the state adopt a *frequency analysis*, which compares the sentence imposed to the sentences imposed in similar cases. There was considerable debate over which cases to include in the analysis: in particular, should all "death eligible" cases be considered, or should the analysis only examine cases for which the death penalty was actually sought? In *Marshall*, the New Jersey Supreme Court temporized and considered both.

In his analysis, Baldus considered several factors when searching for similar cases. First, Robert Marshall was the principal in a contract-murder: in other words, he was the one who hired McKinnon to kill his wife. Second, since Marshall anticipated a life insurance payoff from his wife's murder, he could also be viewed as the hit man in a contract-murder: in other words, he was the one who received money for her death. Third, this was a highly premeditated murder of a defenseless wife. Finally, Baldus compared the Marshall case to other murders with a premeditated robbery, pecuniary motive, and defenseless victim.

For example, there were six other cases involving contract-murder for which the prosecutor sought the death penalty, and three additional cases for which the prosecutor did not seek the death penalty. Only two defendants actually received a death sentence at the penalty trial. We might take this to mean that contract-murder is unlikely to result in the death penalty, and view Marshall's sentence as disproportionate on that basis.

Unfortunately, there is a problem with this approach. If we compare Marshall's case to other contract murders, we find that 2 out of 9 defendants in similar cases received the death penalty. If we take this to mean that Marshall's death penalty is disproportionate, then the next defendant in a similar case can argue that only 2 out of *10* defendants in similar cases received the death penalty. If the death penalty was disproportionate in

Marshall's case, it would be disproportionate in all later cases as well. By the same argument, holding that the death penalty was proportionate in one case would make it proportionate in all similar cases. This effect, known by scientists as *positive feedback*, makes it virtually impossible to alter a trend.

As a result, even though the New Jersey Supreme Court continued to use frequency analysis and mathematical models in the proportionality review of death penalty cases, not one death sentence was reversed on the basis of this analysis.

The Royal Book of Oz

One of the biggest problems with trying to perform a proportionality review is to find comparable cases. In theory, this is easy: the law is based on precedent, so when the New Jersey Supreme Court upheld Marshall's death sentence, it established that in all similar cases, the death sentence would be appropriate.

The problem is defining a similar case. For example, one of the cases compared to Marshall's was that of Michael Rose, an intellectually disabled man with no prior record. Zoran Cveticanin paid Rose $540 to murder his stepmother, Kathryn Cveticanin, who was pregnant and stood to inherit his father's money. Rose stabbed her multiple times before beating her to death with a blunt object. The prosecutor sought the death penalty, though the penalty trial returned a sentence of life in prison with a 30-year parole ineligibility period. We may view this outcome as a judgment by society that the death penalty would have been disproportionate. Zoran fled to Yugoslavia, where he was tried, convicted, and sentenced to 30 years' hard labor.

The path taken by Florida and a number of other states suggests an alternative way to evaluate the appropriateness of the death penalty. As of 2013, Florida law provided a list of 16 aggravating and 7 specific mitigating factors (as well as an eighth catch-all category). Based on the balance between mitigating and aggravating factors, the law concludes whether the death penalty is proportionate or not. If the Rose case had been tried in Florida, the judge would weigh the mitigating factors against the aggravating factors. The facts that Rose was mentally disabled, had no prior record, and was under the substantial domination of Zoran Cveticanin, would weigh in his favor, but the fact that Rose was paid for a particularly brutal murder would weigh against him.

It is unclear how to balance the aggravating and mitigating factors. In this case, the mitigating factors outnumbered the aggravating factors, but we might be uncomfortable with such a simple equation: would the murder of a child (one aggravating factor) be mitigated by lack of a prior criminal history and impairment due to alcohol (two mitigating factors)? What if the victim was a police officer (one aggravating factor substituted for another), or the murderer had a criminal history but was mentally disabled (one mitigating factor substituted for another mitigating factor)? To answer this question, we might consider what seems to be a completely unrelated problem: Who wrote *The Royal Book of Oz*?

In 1900, Lyman Frank Baum published *The Wonderful Wizard of Oz*. The book proved so popular that he wrote 13 sequels before suffering a stroke and dying in 1919. The publishers, not wanting to give up a financial gold mine, contracted Ruth Plumly Thompson to write new *Oz* books. The fifteenth book in the series, *The Royal Book of Oz*, appeared in 1921. Baum was credited as author with Thompson as editor.

Thompson eventually wrote 20 more *Oz* books under her own name. However, there has always been controversy over the authorship of *The Royal Book of Oz*. Many literary scholars believe that *The Royal Book of Oz* was entirely Thompson's creation. This leads us to a situation akin to that of deciding whether the death sentence is appropriate for a particular offender. With the *Oz* books, we have many narratives with interacting characters, and two regions which we might label "Written by Baum" and "Written by Thompson." The question at hand is where one particular narrative should be placed. With capital cases, we have many narratives with interacting characters, and two regions which we might label "Deathworthy" and "Not deathworthy." The question at hand is where one particular narrative should be placed.

In both cases, a non-mathematical approach is to consider the similarities and differences between the narratives, and try to place them in the appropriate regions. But in 2003, José Nilo G. Binongo applied mathematics to the authorship question. To do this, Binongo took each of the novels and coded them according to a certain scheme. There are a number of ways of doing so, though the most helpful focused on the use of 50 *function words* that appear very commonly in English texts: these are words like "the," "and," "of," and so on. In this way, each book can be converted into a *frequency vector*, which consists of an ordered sequence of numbers that corresponds to how often a particular function word occurs.

For example, we might take the terms of our frequency vector to correspond to how often the words "the," "and," "of," "with," "or," and "to" occur in the text, and a frequency vector like <0.067, 0.037, 0.021, 0.007, 0.002, 0.026> would indicate that 0.067 (6.7%) of the words in the text were "the," 3.7% were "and," 2.1% were "of," and so on. This frequency vector can be viewed as the address of a book. We can then find the addresses of other books, some of which were written by Baum, others of which were written by Thompson.

Binongo then applied *principal component analysis* (PCA) to the dataset. The following analogy might be helpful in understanding principal component analysis. Suppose you're in a crowd of people at a fair. If you look into the crowd, it seems that there is no rhyme or reason to how people are distributed.

However, suppose you could look at the crowd from above. From this viewpoint, you would see that people were not randomly distributed about the fairgrounds, but that there were distinct *clusters* of people: around a popular food vendor, for example. Principal component analysis is a mathematical method of identifying the best place to view the crowd so that the clusters can be easily identified.

When Binongo applied PCA to the *Oz* books, he found that they could be cleanly separated into two clusters: one included all the books definitely written by Baum, and the other included all the books definitely written by Thompson. Moreover, the fifteenth book, whose authorship was unknown, fell clearly into the region of Thompson's books. Thus, from a mathematical point of view, *The Royal Book of Oz* is more like a Thompson book than a Baum book.

There's no reason why we can't do the same thing with capital punishment cases. Each case is different, just as each novel is different. But just as we can specify the address of a book using a set of function words, we can specify the address of a case using a set of characteristics. For example, we might use the aggravating and mitigating factors listed by Florida, so a vector like <1, 0, 0, 1, 1, . . . > would indicate by 1's and 0's whether a factor is present or absent: here the first factor is present, the second and third are absent, the fourth and fifth are present, and so on.

An alternative to using actual death penalty cases is to consider "evolving standards of decency," a notion alluded to in *Woodson v. North Carolina* (1976). In effect, if polls show support for the death penalty, then it is not "cruel and unusual" within the meaning of the 8th Amendment. The problem is that polls are a very blunt instrument for measuring a very complex topic. Someone who supports capital punishment in the abstract might never come across a case where he or she would consider it appropriate; others who reject capital punishment might reconsider this stance when confronted by a Ted Bundy or an Osama bin Laden.

PCA offers a much finer instrument. Imagine that we could present respondents with narratives and ask them whether they feel the death sentence would be appropriate for the case. As with Binongo's analysis of the *Oz* canon, these narratives can then be converted into a vector and placed in a high-dimensional space; PCA can then be used to draw a line between deathworthy cases and non-deathworthy cases; and community standards can then be used to determine the appropriate sentence for any actual case.

The Cost of Death

But even if community standards support the death penalty, is it rational to impose it? Again, we can perform a cost-benefit analysis. Since the execution of a criminal returns nothing of value to society, most advocates appeal to the deterrent effect of capital punishment. A report cited by the *Wall Street Journal* suggested that each execution prevents 75 or more murders. This might seem to make the death penalty worthwhile. But this figure misses a key point: even if we grant that the figure is correct, the cost of executing a single prisoner is considerable; could that money, spent in other areas of law enforcement, prevent *more* than 75 murders?

These costs became an important consideration in New Jersey, which reinstituted the death penalty in 1982, but as of 2006 had yet to execute a single prisoner. In that year, it appointed a commission to study the death penalty, particularly in comparison to life imprisonment without the possibility of parole. Among other things, the New Jersey Death Penalty Commission examined the cost of execution, and made the following observations.

1. The Office of the Public Defender estimated that eliminating the death penalty would save it $1.46 million each year, based on the additional workload generated by capital cases.
2. The Department of Corrections estimated that replacing the death penalty with a sentence of life without the possibility of parole would save about $1 million per prisoner over the lifetime of the prisoner, because housing a prisoner on death row costs about $32,000 per year more than housing a prisoner in the general population.
3. The Administrative Office of the Courts estimated that each proportionality review costs about $100,000.

The commission concluded that the cost of execution was significantly higher than the cost of life imprisonment without the possibility of parole. While the commission declined to give an exact number, noting that it was impossible to form a precise estimate, others estimated that replacing the death penalty with a sentence of life without the possibility of parole would save the state $11 million each year. As a result of the commission's report, New Jersey abolished the death penalty on November 17, 2007. Although there were a number of factors involved, the high cost of execution had to weigh heavily on the state, which was facing record deficits and the highest unemployment rate in 32 years.

A similar report in Maryland showed that the state had spent $186 million between 1978 and 2008—to execute just five individuals. Again, the crucial question is that even if the death penalty deters, does it deter $186 million worth? In 2013, Maryland joined the ranks of states that had abolished the death penalty; although the high cost of the death penalty was not cited as a reason for its abolition, it's not unlikely that it weighed heavily on the minds of legislators.

Meanwhile, a 2011 report estimated that California spent more than four billion dollars since 1978 to execute just 13 persons. If we accept that each execution prevents 75 murders, then the death penalty in California prevents about 30 murders a year—at a cost of about four million dollars per murder prevented. As with three-strikes, we might ask: Could the money have been better spent?

The state itself estimated that replacing the death penalty with life without the possibility of parole would save $130 million each year, enough to hire about 1,000 additional law enforcement officers. If each of these officers prevented a single murder in their entire law enforcement *career*, either through their presence on the streets (the patrol model) or the deterrent effect of greater law enforcement resources (the warrant model), they would outperform the death penalty in murders prevented.

In November 2012, California voters considered Proposition 34, which would repeal the death penalty and replace it with life imprisonment without the possibility of parole. It failed—but 48% of the voters supported the abolition of the death penalty, giving hope to those who seek to abolish it.

The Constitutional Equation

> We hold these truths to be self-evident ...
>
> —THOMAS JEFFERSON, *THE DECLARATION OF INDEPENDENCE* (1776)

In the preceding chapters, we've seen how mathematics can be used to support constitutional imperatives, and assess legislation before and after implementation. But mathematics alone is ineffective: for better or for worse, society is created by lawmakers and politicians.

To that end, we summarize our conclusions, grounded in mathematics, about the institutions created by the Constitution. In some cases, mathematics suggests that the institutions, as they exist, are consistent with our concepts of justice and fairness. In other cases, mathematics suggests a better alternative.

The Census

The Census Bureau should be permitted to use sampling methods when conducting the census, even for apportionment purposes. However, this will require a change in the apportionment process, as discussed next.

Apportionment

Replace the priority values for apportionment with priority intervals, based on statistical confidence intervals for a state's population. As this may result in ties, where two or more seats are assigned simultaneously, this will require amending the 1911 Apportionment Act to permit more than 435 congressmen.

Representation

Implement weighted voting in the House of Representatives, with the weights of each representative proportional to the square root of the apportionment population of their district.

Voting

Promote approval voting systems for all elections (except referendums, recall votes, and other "yes/no" decisions, where plurality suffices).

Gerrymandering

Require that congressional districts meet several compactness measures, such as the ratio of the district's area to the area of the circumscribing circle (the Reock measure, used by Michigan) or reference to the sum of district perimeters (the method used by Iowa and Colorado). Use partisan symmetry as a way to gauge the extent of partisan gerrymandering, and reconsider the decisions in *Vieth v. Jubelirer* (2004) and *LULAC v. Perry* (2006) that deny the justiciability of partisan gerrymandering.

Probable Cause

Use hit rates as the key method of evaluating a law enforcement policy with regard to whether it satisfies "individualized suspicion" and whether it is being used in a race-neutral manner.

Juries

Implement the 80% rule as a basic measure of jury discrimination, and promote the use of statistical methods to assess the existence of discrimination. Require that juries consist of 12 persons who render a verdict unanimously. Alternatively, consider 18-person juries requiring 16 jurors to agree to a verdict.

Criminal Justice

Repeal "three strikes" and the death penalty and replace them with appropriate sentencing guidelines. Use centrality measures from social network analysis to evaluate the role a person plays in a conspiracy.

Select Topical Bibliography

1.21. STAND UP AND BE ESTIMATED

Anderson, M., Fienberg, S. E. "To Sample or Not to Sample? The 2000 Census Controversy." *Journal of Interdisciplinary History*, xxx:1 (Summer 1999), 1–36.

Anderson, M., Fienberg, S. E. "Census 2000 and the Politics of Census Taking." *Society*, Nov./Dec. 2001, pp. 17–25.

Barbara, V. P., Mason, R. O., Mitroff, I. I. "Federal Statistics in a Complex Environment: The Case of the 1980 Census." *American Statistician*, vol. 37, no. 3 (Aug. 1983), pp. 203–212.

Belin, T. R., Rolph, J. E. "Can We Reach Consensus on Census Adjustment?" *Statistical Science*, vol. 9, no. 4 (Nov. 1994), pp. 486–508.

Brunell, T. L. "Science and Politics in the Census." *Society*, Nov./Dec. 2001, pp. 11–16.

Cantwell, P. J., Hogan, H., Styles, K. M. "Imputation, Apportionment, and Statistical Methods in the U.S. Census: Issues Surrounding Utah v. Evans." Statistical Research Division, U.S. Bureau of the Census, report issued March 7, 2005.

Cantwell, P. J., Hogan, H., Styles, K. M. "The Use of Statistical Methods in the U.S. Census: 'Utah v. Evans'." *American Statistician*, vol. 58, no. 3 (Aug. 2004), pp. 203–212.

Dwyer, N. T. "Utah v. Evans: How Census 2000's 'Sampling in Disguise' Fooled the Supreme Court into Allocating Utah's Seat in the U.S. House of Representatives to North Carolina." *Pepperdine Law Review*, vol. 31, issue 4, article 5.

Edmonston, B. "The Case for Modernizing the U.S. Census." *Society*, Nov./Dec. 2001, pp. 43–53.

Edmonston, B., Schultze, C., eds. *Modernizing the U.S. Census*. Washington, D.C., National Academies Press, 1995.

Ericksen, E. P., Kadane, J. B. "Estimating the Population in a Census Year: 1980 and Beyond." *Journal of the American Statistical Association*, vol. 80, no. 389 (Mar. 1985), pp. 98–109.

Freedman, D. A., Wachter, K. W. "On the Likelihood of Improving the Accuracy of the Census through Statistical Adjustment." Lecture Notes–Monograph Series, vol. 40, Statistics and Science: A Festschrift for Terry Speed (2003), pp. 197–230.

Freedman, D. A., Wachter, K. W. "Census Adjustment: Statistical Promise or Illusion?" *Society*, Nov./Dec. 2001, pp. 26–33.

Prewitt, K. "The US Decennial Census: Political Questions, Scientific Answers." *Population and Development Review*, vol. 26, no. 1 (Mar. 2000), pp. 1–16.

Razi, B. J. "Census Politics Revisited: What to Do When the Government Can't Count?" *American University Law Review*, vol. 48 (June 1999), pp. 1101–1138.

Steffey, D. L. *A Review of the Census Undercount Issue.* Faculty Fellows Program, Center for California Studies, California State University, October 1997.

Wright, T. "Census 2000: Who Says Counting Is Easy As 1-2-3?" *Government Information Quarterly,* vol. 17, no. 2 (2000), pp. 121–136.

Wright, T. "A One-Number Census: Some Related History." *Science,* New Series, vol. 283, no. 5401 (Jan. 22, 1999), pp. 491–492.

Wright, T. "Sampling and Census 2000: The Concepts: Established Statistical Methods Can Reduce Net Undercounting of the Population—If They Are Allowed." *American Scientist,* vol. 86, no. 3 (May–June 1998), pp. 245–253.

1.22. (NEARLY) EQUAL REPRESENTATION

Balinski, M. L., Young, H. P. "The Jefferson Method of Apportionment." *SIAM Review,* vol. 20, no. 2 (Apr. 1978), pp. 278–284.

Balinski, M. L., Young, H. P. "On Huntington Methods of Apportionment." *SIAM Journal on Applied Mathematics,* vol. 33, no. 4 (Dec. 1977), pp. 607–618.

Edelman, P. H. "Getting the Math Right: Why California Has Too Many Seats in the House of Representatives." *Vanderbilt Law Review* vol. 59 (March 2006), pp. 297–346.

Edelman, P. H., Sherry, S. "Pick a Number, Any Number: State Representation in Congress after the 2000 Census." *California Law Review,* vol. 90, no. 1 (Jan. 2002), pp. 211–222.

Huntington, E. V. "Methods of Apportionment in Congress." *American Political Science Review,* vol. 25, no. 4 (Nov. 1931), pp. 961–965.

Huntington, E. V. "A New Method of Apportionment of Representatives." *Quarterly Publications of the American Statistical Association,* vol. 17, no. 135 (Sep. 1921), pp. 859–870.

Huntington, E. V. "The Mathematical Theory of the Apportionment of Representatives." *Proceedings of the National Academy of Sciences of the United States of America,* vol. 7, no. 4 (Apr. 15, 1921), pp. 123–127.

McLawhorn, Jr., R. E. "Apportionment or Size? Why the U.S. House of Representatives Should Be Expanded." *Alabama Law Review* vol. 62 (2011), pp. 1069–1091.

1.23. WEIGHTING FOR A FAIR VOTE

Banzhaf III, J. F. "One Man, ? Votes: Mathematical Analysis of Voting Power and Effective Representation." *George Washington Law Review,* vol. 36 (1967–1968), pp. 808–823.

Gelman, A., Katz, J. N., Bafumi, J. "Standard Voting Power Indexes Do Not Work: An Empirical Analysis." *British Journal of Political Science,* vol. 34, no. 4 (Oct. 2004), pp. 657–674.

Gelman, A., Katz, J. N., Tuerlinckx, F. "The Mathematics and Statistics of Voting Power." *Statistical Science* vol. 17, no. 4 (2002), pp. 420–435.

Grofman, B. "Fair Apportionment and the Banzhaf Index." *American Mathematical Monthly,* vol. 88, no. 1 (Jan. 1981), pp. 1–5.

Grofman, B., Scarrow, H. "Weighted Voting in New York." *Legislative Studies Quarterly,* vol. 6, no. 2 (May 1981), pp. 287–304.

Margolis, H. "The Banzhaf Fallacy." *American Journal of Political Science*, vol. 27, no. 2 (May 1983), pp. 321–326.

Merrill III, S. "Approximations to the Banzhaf Index of Voting Power." *American Mathematical Monthly*, vol. 89, no. 2 (Feb. 1982), pp. 108–110.

1.24. THE IMPOSSIBILITY OF DEMOCRACY

Arrow, K. J. "A Difficulty in the Concept of Social Welfare." *Journal of Political Economy*, vol. 58, no. 4 (Aug. 1950), pp. 328–346.

Balinski, M., Laraki, R. *Majority Judgment: Measuring, Ranking, and Electing.* Cambridge, MA: MIT Press, 2011.

Bertelli, A., Richardson Jr., L. E. "Ideological Extremism and Electoral Design. Multimember versus Single Member Districts." *Public Choice*, vol. 137, no. 1/2 (Oct. 2008), pp. 347–368.

Brams, S. J. *Mathematics and Democracy: Designing Better Voting and Fair Division Procedures.* Princeton, NJ: Princeton University Press, 2008.

Brams, S. J., Fishburn, P.C. *Approval Voting*, 2nd ed. New York: Springer-Verlag, 2007.

Brams, S. J., Kilgour, D. M. "Narrowing the Field in Elections: The Next-Two Rule." *Journal of Theoretical Politics*, vol. 24, no. 4 (2012), pp. 507–525.

Cox, G. W. "Centripetal and Centrifugal Incentives in Electoral Systems." *American Journal of Political Science*, vol. 34, no. 4 (Nov. 1990), pp. 903–935.

Dougherty, K. L., Edward, J. *The Calculus of Consent and Constitutional Design.* New York: Springer-Verlag, 2011.

Gill, J., Gainous, J. "Why Does Voting Get So Complicated? A Review of Theories for Analyzing Democratic Participation." *Statistical Science*, vol. 17, no. 4, Voting and Elections (Nov. 2002), pp. 383–404.

Pacelli, A. M., Taylor, A. D. *Mathematics and Politics*, 2nd ed. New York: Springer-Verlag, 2008.

Schweigert, B. J. "'Now for a Clean Sweep!': Smiley v. Holm, Partisan Gerrymandering, and At-Large Congressional Elections." *Michigan Law Review*, vol. 107, no. 1 (Oct. 2008), pp. 133–164.

Simeone, B., Pukelsheim, F., eds. *Mathematics and Democracy.* Heidelberg: Springer-Verlag, 2006.

1.4. DRAGONS AND DUMMYMANDERS

Angel, S., Parent, J. "Non-compactness as Voter Exchange: Towards a Constitutional Cure for Gerrymandering." *Northwestern Interdisciplinary Law Review*, vol. 4, no. 1, 2011, pp. 89–145.

Browning, R. X., King, G. "Seats, Votes, and Gerrymandering: Estimating Representation and Bias in State Legislative Redistricting." *Law & Policy*, vol. 9, no. 3 (July 1987), pp. 305–322.

Calabrese, S. "Multimember District Congressional Elections." *Legislative Studies Quarterly*, vol. 25, no. 4 (Nov. 2000), pp. 611–643.

Case, J. "Flagrant Gerrymandering: Help from the Isoperimetric Theorem?" *SIAM News*, vol. 40, no. 9, Nov. 2007, n.p.

Cox, A. "Partisan Fairness and Redistricting Politics." *New York University Law Review*, June 2004, pp. 751–802.

Friedman, J. N., Holden, R. T. "Optimal Gerrymandering: Sometimes Pack, but Never Crack." *American Economic Review*, vol. 98, no. 1 (Mar. 2008), pp. 113–144.

Fromer, J. C. "An Exercise in Line-Drawing: Deriving and Measuring Fairness in Redistricting." *Georgetown Law Journal*, vol. 93 (June 2005), pp. 1547–1621.

Greene, J. "Judging Partisan Gerrymanders under the Elections Clause." *Yale Law Journal*, vol. 114, no. 5 (Mar. 2005), pp. 1021–1062.

Grofman, B., Brunell, T. "The Art of the Dummymander: The Impact of Recent Redistrictings on the Partisan Makeup of Southern House Seats." In *Redistricting in the New Millennium,* ed. P. F. Galderisi. Washington D.C.: Lexington Books, 2005.

Grofman, B., King, G. "The Future of Partisan Symmetry as a Judicial Test for Partisan Gerrymandering after *LULAC v. Perry.*" *Election Law Journal*, vol. 6, no. 1 (2007), pp. 2–35

Issacharoff, S., Karlan, P. S. "Where to Draw the Line?: Judicial Review of Political Gerrymanders." *University of Pennsylvania Law Review*, vol. 153, no. 1, Symposium: The Law of Democracy (Nov. 2004), pp. 541–578.

Kang, M. S. "When Courts Won't Make Law: Partisan Gerrymandering and a Structural Approach to the Law of Democracy." *Ohio State Law Journal*, vol. 68:1097 (2007).

King, G., Browning, R. X. "Democratic Representation and Partisan Bias in Congressional Elections." *American Political Science Review*, vol. 81, no. 4 (Dec. 1987), pp. 1251–1273.

Niemi, R., Grofman, B., Carlucci, C., Hofeller, T. "Measuring Compactness and the Role of a Compactness Standard in a Test for Partisan and Racial Gerrymandering." *Journal of Politics*, vol. 52, no. 4 (Nov. 1990), pp. 1155–1181.

Niemi, R. G., Deegan, Jr., J. "A Theory of Political Districting." *American Political Science Review*, vol. 72, no. 4 (Dec. 1978), pp. 1304–1323.

Niemi, R. G., Fett, P. "The Swing Ratio: An Explanation and an Assessment." *Legislative Studies Quarterly*, vol. 11, no. 1 (Feb. 1986), pp. 75–90.

Owen, G., Grofman, B. "Optimal Partisan Gerrymandering." *Political Geography Quarterly*, vol. 7, no. 1 (January 1988), pp. 1–22.

Pildes, R. H., Niemi, R. G. "Expressive Harms, 'Bizarre Districts,' and Voting Rights: Evaluating Election-District Appearances after Shaw v. Reno." *Michigan Law Review*, vol. 92, no. 3 (Dec. 1993), pp. 483–587.

Polsby, D. D., Popper, R. D. "The Third Criterion: Compactness as a Procedural Safeguard against Partisan Gerrymandering." *Yale Law & Policy Review*, vol. 9, no. 2 (1991), pp. 301–353.

Taylor, P. J. "A New Shape Measure for Evaluating Electoral District Patterns." *American Political Science Review*, vol. 67, no. 3 (Sep. 1973), pp. 947–950.

Young, H. P. "Measuring the Compactness of Legislative Districts." *Legislative Studies Quarterly*, vol. 13, no. 1 (Feb. 1988), pp. 105–115.

2.1. THE WORST WAY TO ELECT A PRESIDENT, EXCEPT FOR ALL THE REST

Fischer, A. J. "The Probability of Being Decisive." *Public Choice*, vol. 101, no. 3/4 (Oct. 1999), pp. 267–283.

Gelman, A., King, G., Boscardin, W. J. "Estimating the Probability of Events That Have Never Occurred: When Is Your Vote Decisive?" *Journal of the American Statistical Association*, vol. 93, no. 441 (Mar. 1998), pp. 1–9.

Gringer, D. "Why the National Popular Vote Plan Is the Wrong Way to Abolish the Electoral College." *Columbia Law Review*, vol. 108 (January 2008), pp. 182–230.

Grofman, B., Brunell, T. "Distinguishing Between the Effects of Swing Ratio and Bias on Outcomes in the U.S. Electoral College, 1900–1992." *Electoral Studies*, vol. 16, no. 4 (1997), pp. 471–487.

Margolis, H. "Probability of a Tie Election." *Public Choice*, vol. 31 (Fall 1977), pp. 135–138.

Mayer, L. S., Good, I. J. "Is Minimax Regret Applicable to Voting Decisions?" *American Political Science Review*, vol. 69, no. 3 (Sep. 1975), pp. 916–917.

Mulligan, C. B., Hunter, C. G. "The Empirical Frequency of a Pivotal Vote." *Public Choice*, vol. 116, no. 1/2 (Jul. 2003), pp. 31–54.

Natapoff, A. "A Mathematical One-Man One-Vote Rationale for Madisonian Presidential Voting Based on Maximum Individual Voting Power." *Public Choice*, vol. 88, no. 3/4 (1996), pp. 259–273.

Thomas, A. C., Gelman, A., King, G., Katz, J. N. "Estimating Partisan Bias of the Electoral College Under Proposed Changes in Elector Apportionment." *Statistics, Politics, and Policy*, vol. 4, no. 1 (2013), pp. 1–13.

A4.1. STOP AND FRISK

Gelman A., Fagan, J., Kiss, A. "An Analysis of the New York City Police Department's 'Stop-and-Frisk' Policy in the Context of Claims of Racial Bias." *Journal of the American Statistical Association*, vol. 102, no. 479 (Sep. 2007), pp. 813–823.

Glaser, J. "The Efficacy and Effect of Racial Profiling: A Mathematical Simulation Approach." *Journal of Policy Analysis and Management*, vol. 25, no. 2 (Spring 2006), pp. 395–416.

Harcourt, B. E. "Rethinking Racial Profiling: A Critique of the Economics, Civil Liberties, and Constitutional Literature, and of Criminal Profiling More Generally." *University of Chicago Law Review*, vol. 71, no. 4 (Autumn 2004), pp. 1275–1381.

Kaminsky, P. L. "The Wrap on Probable Cause: The Fourth Amendment Contained." *St. Thomas Law Review*, vol. 6 (1994), pp. 449–477.

McCauliff, C.M.A. "Burdens of Proof: Degrees of Belief, Quanta of Evidence, or Constitutional Guarantees?" *Vanderbilt Law Review*, vol. 35 (November 1982), pp. 1293–1335.

O'Neill, T. P. "Vagrants in Volvos: Ending Pretextual Traffic Stops and Consent Searches of Vehicles in Illinois." *Loyola University Chicago Law Journal*, vol. 40 (Summer 2009), pp. 745–779.

Taslitz, A. E. "Racial Auditors and the Fourth Amendment: Data with the Power to Inspire Political Action." *Law and Contemporary Problems*, vol. 66, no. 3, The New Data: Over-Representation of Minorities in the Criminal Justice System (Summer 2003), pp. 221–298.

A4.2. REVEREND THOMAS BAYES AND THE LAW

Finkelstein, M. O., Fairley, W. B. "A Bayesian Approach to Identification Evidence." *Harvard Law Review*, vol. 83, no. 3 (Jan. 1970), pp. 489–517.

Harcourt, B. E., Meares, T. L. "Randomization and the Fourth Amendment." *University of Chicago Law Review*, vol. 78, no. 3 (Summer 2011), pp. 809–877.

Hardin, R. "Civil Liberties in the Era of Mass Terrorism." *Journal of Ethics*, vol. 8, no. 1, Terrorism (2004), pp. 77–95.

Kaye, D. H. "Trawling DNA Databases for Partial Matches: What Is the FBI Afraid Of?" *Cornell Journal of Law and Public Policy*, 19 (Fall 2009), pp. 145–171.

Liebman, J. S., Blackburn, S., Mattern, D., Waisnor, J., "The Evidence of Things Not Seen: Non-Matches as Evidence of Innocence." *Iowa Law Review* vol. 98 (Jan. 2013), pp. 577–688.

Luna, E. "The Bin Laden Exception." *Northwestern University Law Review Colloquy*, vol. 106 (Feb. 2012), pp. 230–247.

Roman-Santos, C. "Concerns Associated with Expanding DNA Databases." *Hastings Science & Technology Law Journal*, vol. 2 (Summer 2010), pp. 267–299.

Rubinstein, I. S., Lee, R. D., Schwartz, P. M. "Data Mining and Internet Profiling: Emerging Regulatory and Technological Approaches." *University of Chicago Law Review*, vol. 75, no. 1 (Winter 2008), pp. 261–285.

Slobogin, C. "Government Data Mining and the Fourth Amendment." *University of Chicago Law Review*, vol. 75 (Winter 2008), pp. 317–341.

Tverdek, E. "The Limits to Terrorist Profiling." *Public Affairs Quarterly*, vol. 20, no. 2 (Apr. 2006), pp. 175–203.

A5. "THE MAN OF STATISTICS"

Finkelstein, M. O. "The Application of Statistical Decision Theory to the Jury Discrimination Cases." *Harvard Law Review*, vol. 80, no. 2 (Dec. 1966), pp. 338–376.

Kaye, D. "Statistical Evidence of Discrimination." *Journal of the American Statistical Association*, vol. 77, no. 380 (Dec. 1982), pp. 773–783.

King, A. D. "'Gross Statistical Disparities' as Evidence of a Pattern and Practice of Discrimination: Statistical versus Legal Significance." 22 *Labor Lawyer* 271 (2007), pp. 271–292.

Starkey, B. S. "Criminal Procedure, Jury Discrimination & the Pre-Davis Intent Doctrine: The Seeds of a Weak Equal Protection Clause." *American Journal of Criminal Law*, vol. 38 (Fall 2010), pp. 1–48.

Sugrue, T. J., Fairley, W. B. "A Case of Unexamined Assumptions: The Use and Misuse of the Statistical Analysis of Castaneda/Hazelwood in Discrimination Litigation." 24 *Boston College Law Review*, 925 (1983), pp. 925–960.

A6.1. DESPAIR OVER DISPARITY

Detre, P. A. "A Proposal for Measuring Underrepresentation in the Composition of the Jury Wheel." *Yale Law Journal*, vol. 103, no. 7 (May 1994), pp. 1913–1938.

Dickson, D. "State Court Defiance and the Limits of Supreme Court Authority: Williams v. Georgia Revisited." *Yale Law Journal,* vol. 103, no. 6 (Apr. 1994), pp. 1423–1481.

Meier, P., Sacks, J., Zabell, S. L. "What Happened in Hazelwood: Statistics, Employment Discrimination, and the 80% Rule." *American Bar Foundation Research Journal,* vol. 9, no. 1 (Winter 1984), pp. 139–186.

A6.2. ONCE IS AN ACCIDENT . . .

Barrett, B. E. "Detecting Bias in Jury Selection." *American Statistician,* vol. 61, no. 4 (Nov. 2007), pp. 296–301.

Kairys, D., Kadane, J. B. "Fair Numbers of Peremptory Challenges in Jury Trials." *Journal of the American Statistical Association,* vol. 74, no. 368 (Dec. 1979), pp. 747–753.

Franklin, L. A. "Bayes' Theorem, Binominal Probabilities, and Fair Numbers of Peremptory Challenges in Jury Trials." *College Mathematics Journal,* vol. 18, no. 4 (Sep. 1987), pp. 291–299.

Page, A. "Batson's Blind-Spot: Unconscious Stereotyping and the Peremptory Challenge." *Boston University Law Review,* 85 (Feb. 2005), pp. 155–262.

A6.3. ~~12 6 5 10~~ *N*-ANGRY MEN

Devine, D. J., Clayton, L. D., Dunford, B. B., Seying, R., Pryce, J. "Jury Decision Making: 45 Years of Empirical Research on Deliberating Groups." *Psychology, Public Policy and Law,* vol. 7 (Sep. 2001), pp. 622–727.

Fabian, V. "On the Effect of Jury Size." *Journal of the American Statistical Association,* vol. 72, no. 359 (Sep. 1977), pp. 535–536.

Friedman, H. "Trial by Jury: Criteria for Convictions, Jury Size and Type I and Type II Errors." *American Statistician,* vol. 26, no. 2 (Apr. 1972), pp. 21–23.

Garvey, S. P., Hannaford-Agor, P., Hans, V. P., Mott, N. L., Munsterman, G. T., Wells, M. T. "Juror First Votes in Criminal Trials." *Journal of Empirical Legal Studies,* vol. 1, issue 2 (July 2004), pp. 371–398.

Gelfand, A. E., Solomon, H. "Analyzing the Decision-Making Process of the American Jury." *Journal of the American Statistical Association,* vol. 70, no. 350 (Jun. 1975), pp. 305–310.

Gelfand, A. E., Solomon, H. "Modeling Jury Verdicts in the American Legal System." *Journal of the American Statistical Association,* vol. 69, no. 345 (Mar. 1974), pp. 32–37.

Hans, V. P., Vidmar, N. "The *American Jury* at Twenty-Five Years." *Law and Social Inquiry,* vol. 16, no. 2 (Spring 1991), pp. 325–351.

Helland, E., Raviv, Y. "The Optimal Jury Size When Jury Deliberation Follows a Random Walk." *Public Choice* vol. 134 (2008), pp. 255–262.

Kalven Jr., H., Zeisel, H. *The American Jury.* Boston: Little, Brown, 1966.

Kaye, D. "And Then There Were Twelve: Statistical Reasoning, the Supreme Court, and the Size of the Jury." *California Law Review,* vol. 68, no. 5 (Sep. 1980), pp. 1004–1043.

Klevorick A. K., Rothschild, M. "A Model of the Jury Decision Process." *Journal of Legal Studies*, vol. 8, no. 1 (Jan. 1979), pp. 141–164.

Nagel, S. S., Neef, M. "Deductive Modeling to Determine an Optimum Jury Size and Fraction Required to Convict." *Washington University Law Review*, vol. 1975, issue 4, 1975.

Reagan, R. T. "Supreme Court Decisions and Probability Theory: Getting the Analysis Right." *University of Detroit Mercy Law Review*, vol. 77 (Summer 2000), pp. 835–873.

Saks, M. J., Martin, M. W. "A Meta-Analysis of the Effects of Jury Size." *Law and Human Behavior*, vol. 21, no. 5 (Oct. 1997), pp. 451–467.

Sandys, M., Dillehay, R. C. "First-Ballot Votes, Predeliberation Dispositions, and Final Verdicts in Jury Trials." *Law and Human Behavior*, vol. 19, no. 2 (Apr. 1995), pp. 175–195.

Smith, A., Saks, M. "The Case for Overturning *Williams v. Florida* and the Six-Person Jury: History, Law, and Empirical Evidence." *Florida Law Review*, vol. 60 (April 2008), pp. 441–471.

Sperlich, P. W. "Trial by Jury: It May Have a Future." *Supreme Court Review*, vol. 1978 (1978), pp. 191–224.

Thomas III, G. C., Pollack, B. S. "Rethinking Guilt, Juries, and Jeopardy." *Michigan Law Review*, vol. 91, no. 1 (Oct. 1992), pp. 1–33.

A8.1. THE PERIL AND PROMISE OF SOCIAL NETWORK ANALYSIS

Alston, P. "The CIA and Targeted Killings Beyond Borders." *Harvard National Security Journal*, vol. 2 (2011), pp. 283–446.

Baker, W. E., Faulkner, R. R. "The Social Organization of Conspiracy: Illegal Networks in the Heavy Electrical Equipment Industry." *American Sociological Review*, vol. 58, no. 6 (Dec. 1993), pp. 837–860.

Borgatti, S. P. "Identifying Sets of Key Players in a Social Network." *Computational and Mathematical Organization Theory*, vol. 12 (2006), pp. 21–34.

Freeman, L. C. "Centrality in Social Networks: Conceptual Clarification." *Social Networks*, vol. 1 (1978/79), pp. 215–239.

Freeman, L. C. "A Set of Measures of Centrality Based on Betweenness." *Sociometry*, vol. 40, no. 1 (Mar. 1977), pp. 35–413.

Hutchins, C. E., Benham-Hutchins, M. "Hiding in Plain Sight: Criminal Network Analysis." *Computational Mathematics Organization Theory*, vol. 16 (2010), pp. 89–111.

Koschade, S. "A Social Network Analysis of Jemaah Islamiyah: The Applications to Counterterrorism and Intelligence." *Studies in Conflict & Terrorism*, vol. 29 (2006), pp. 559–575.

Krebs, V. E. "Mapping Networks of Terrorist Cells." *Connections*, vol. 24, no. 3 (2002), pp. 43–52.

McGloin, J. M. "Policy and Intervention Considerations of a Network Analysis of Street Gangs." *Criminology and Public Policy*, vol. 4, no. 3 (2005), pp. 607–636.

Morselli, C., Masias, V. H., Crespo, F. Laegnle, S. "Predicting Sentencing Outcomes with Centrality Measures." *Security Informatics*, vol. 2, no. 4 (2013), n.p.

Natarajan, M. "Understanding the Structure of a Drug Trafficking Network: A Conversational Analysis." *Crime Prevention Studies*, vol. 11 (2000), pp. 273–298.

Rhodes, C. J., Keefe, E.M.J. "Social Network Topology: A Bayesian Approach." *Journal of the Operational Research Society*, vol. 58, no. 12 (Dec. 2007), pp. 1605–1611.

Schwartz, D. M., Rouselle, T. "Using Social Network Analysis to Target Criminal Networks." *Trends in Organized Crime*, vol. 12 (2009), pp. 188–207.

Strahilevitz, L. J. "A Social Networks Theory of Privacy." *University of Chicago Law Review*, vol. 72 (Summer 2005), pp. 919–988.

Strandburg, K. "Freedom of Association in a Networked World: First Amendment Regulation of Relational Surveillance." *Boston College Law Review*, vol. 49 (May 2008), pp. 741–821.

van der Hulst, R. C. "Introduction to Social Network Analysis (SNA) as an Investigative Tool." *Trends in Organized Crime*, vol. 12 (2009), pp. 101–121.

A8.2. THREE STRIKES FOR THREE STRIKES

Bartolucci, F., Pennoni, F., Francis, B. "A Latent Markov Model for Detecting Patterns of Criminal Activity." *Journal of the Royal Statistical Society*. Series A (Statistics in Society), vol. 170, no. 1 (2007), pp. 115–132.

Berk, R., Bleich, J. "Forecasts of Violence to Inform Sentencing Decisions." *Journal of Quantitative Criminiology*, vol. 30, no. 1 (March 2014), pp. 79–96..

Chemerinsky, E. "Cruel and Unusual: The Story of Lenadro Andrade." *Drake Law Review*, vol. 52 (2003), pp. 1–24.

Cohen, M. A., Piquero, A. R. "New Evidence on the Monetary Value of Saving a High Risk Youth." *Journal of Quantitative Criminology*, vol. 25 (2009), pp. 25–49.

Di Tella, R., Schargrodsky, E. "Do Police Reduce Crime? Estimates Using the Allocation of Police Forces after a Terrorist Attack." *American Economic Review*, vol. 94, no. 1 (Mar. 2004), pp. 115–133.

Gerst, S. A. "Comparative Proportionality Analysis: Its Feasibility and Usefulness in Sentencing." *Phoenix Law Review*, vol. 1 (Spring 2008), pp. 59–115.

Helland, E., Tabarrok, A. "Does Three Strikes Deter? A Nonparametric Estimation." *Journal of Human Resources*, vol. 42, no. 2 (Spring 2007), pp. 309–330.

Marvell, T. B., Moody, C. E. "The Lethal Effects of Three-Strikes Laws." *Journal of Legal Studies*, vol. 30, no. 1 (Jan. 2001), pp. 89–106.

Miller, T. R., Cohen, M. A., Wiersema, B. *Victim Costs and Consequences: A New Look*. Washington, DC: National Institute of Justice, January 1996.

Netter, B. "Using Group Statistics to Sentence Individual Criminals: An Ethical and Statistical Critique of the Virginia Risk Assessment Program." *Journal of Criminal Law and Criminology* (1973–), vol. 97, no. 3 (Spring 2007), pp. 699–729.

Peterson, M., Braiker, H., Polich, S. "Doing Crime: A Survey of California Prison Inmates." RAND Corporation Report, April 1980.

Piquero, A. R., Weisburd, D., eds. *Handbook of Quantitative Criminology*, New York: Springer-Verlag, 2010.

Schild, U. J. "Criminal Sentencing and Intelligent Decision Support." *Artificial Intelligence and Law*, vol. 6 (1998), pp. 151–202.

Schrodt, P. A. "Forecasting Conflict in the Balkans Using Hidden Markov Models." In *Programming for Peace*, ed. R. Trappl. Dordrecht, The Netherlands: Springer-Verlag, 2006.

Schwartz, C. W. "Eighth Amendment Proportionality Analysis and the Compelling Case of William Rummel." *Journal of Criminal Law and Criminology* (1973–), vol. 71, no. 4 (Winter 1980), pp. 378–420.

Shepherd, J. M. "Fear of the First Strike: The Full Deterrent Effect of California's Two- and Three-Strikes Legislation." *Journal of Legal Studies*, vol. 31, no. 1 (Jan. 2002), pp. 159–201.

A8.3. THE PRICE OF PUNISHMENT

Baldus, D. C. *Death Penalty Proportionality Review Project: Final Report to the New Jersey Supreme Court.* September 24, 1991.

Baldus, D.C., Pulaski, C., Woodworth, G. "Comparative Review of Death Sentences: An Empirical Study of the Georgia Experience." *Journal of Criminal Law and Criminology* (1973–), vol. 74, no. 3 (Autumn 1983), pp. 661–753.

Baldus, D. C., Pulaski Jr., C. A., Woodworth, G., Kyle, F. "Identifying Comparatively Excessive Sentences of Death: A Quantitative Approach." *Stanford Law Review*, vol. 33, no. 1 (Nov. 1980), pp. 1–74.

Binongo, J.N.G. "Who Wrote the 15th Book of Oz?" *Chance*, vol. 16, no. 2 (2003), pp. 9–17.

Kaufman-Osborn, T. V., "Proportionality Review and the Death Penalty." *Justice System Journal*, vol. 29, no. 3 (2008), pp. 257–272.

Latzer, B. "The Failure of Comparative Proportionality Review of Capital Cases (with Lessons from New Jersey)." *Albany Law Review*, vol. 64 (2001), pp. 1161–1244.

Mandery, E. J. "In Defense of Specific Proportionality Review." *Albany Law Review*, vol. 65 (2002), pp. 883–934.

Romano, M. "Striking Back: Using Death Penalty Cases to Fight Disproportionate Sentences Imposed Under California's Three Strikes Law." *Stanford Law & Policy Review*, vol. 21 (2010), pp. 311–348.

Sprenger, S. M. "A Critical Evaluation of State Supreme Court Proportionality Review in Death Sentence Cases." *Iowa Law Review*, vol. 73 (March 1988), pp. 719–741.

Walton, D. "Similarity, Precedent and Argument from Analogy." *Artificial Intelligence in the Law*, vol. 18 (2010), pp. 217–246.

Index